Braiding Pultrusion Technology

Braiding Pultrusion Technology

For Advanced Structural Composites

Garry E. Freger
Vladimir N. Kestelman
Dmitry G. Freger

McGraw-Hill

New York Chicago San Francisco Lisbon London
Madrid Mexico City Milan New Delhi San Juan
Seoul Singapore Sydney Toronto

The McGraw-Hill Companies

Cataloging-in-Publication Data is on file with the Library of Congress.

1 2 3 4 5 6 7 8 9 0 DOC/DOC 0 1 0 9 8 7 6 5 4

ISBN 0-07-143478-X

The sponsoring editor for this book was Kenneth P. McCombs and the production supervisor was Sherri Souffrance. It was set in Century Schoolbook by International Typesetting and Composition. The art director for the cover was Margaret Webster-Shapiro.

Printed and bound by RR Donnelley.

This book is printed on recycled, acid-free paper containing a minimum of 50% recycled, de-inked fiber.

McGraw-Hill books are available at special quantity discounts to use as premiums and sales promotions, or for use in corporate training programs. For more information, please write to the Director of Special Sales, McGraw-Hill Professional, Two Penn Plaza, New York, NY 10121-2298. Or contact your local bookstore.

Contents

Preface

The problems of braiding pultrusion of rod-like products based on polymer composites are expounded in this book. The kinematic parameters of braiding are defined using a specific procedure, and the primary dependencies for computation of structural and geometric parameters of materials are described. Rational time and temperature regimes of pultrusion are reported, their effects on physicomechanical properties of materials are estimated, and force parameters of the combined process are analyzed. The production technology for articles of complex configurations broached with reinforcing materials is considered. A combined braiding pultrusion process used in the production of dead-end rods used in excavation is described.

Composite materials used in various intricately shaped structures necessitate application of specific stiffening members, such as rods, beams, braces, frames, longerons, strings, and so on. Because of the high operation loads the elements must bear, their reinforcement schemes should ensure perfect transversal characteristics under various types of loading. To ensure that the amount of these stiffening elements in force structures is large enough, it is strongly recommended that highly productive methods of molding, e.g., braiding or pultrusion, be employed. The former, braiding, creates the required material structure that withstands the needed loads, while the latter, pultrusion, provides a continuous effective process yielding high-quality, accurate products. When united together in a single technological process, these methods combine all their positive features, achieving a spatial reinforcement scheme that is molded continuously into long-length goods. Modern procedures that combine the formation of braided structures with their profile molding by pultrusion produce only a limited range of goods. In addition, no comprehensive theoretical investigations can yet serve as the basis for designing complex-configuration rods for use in force applications, with considerations for operation conditions and specifics of the braiding pultrusion technology. One of the variants of manufacturing complex-configuration products by continuous methods is their broaching, which lessens the stress concentration in critical section sites and elevates the article-bearing capacity. A profile can be broached by short rods, by composite or metal clips, or by fibrous threads or plaits—all depending on the article dimensions and the required mechanical characteristics of the pierced layer. Named technological methods have been developed only as individual processes and have not yet been used in unified technological schedules.

In this respect, the problem remains of designing a single technological procedure that combines the methods described above. Apart from the currently unavailable theoretical and experimental bases, the properties of the materials obtained by this method must be examined in order to study the structural and geometric limitations as well as the technological parameters of molding.

The purpose of the present book is to fill in the gaps described above, to suggest both engineering and design solutions, and to substantiate production regimes of the process.

Chapter 1 of this book deals with the manufacture of composite parts, including braiding and pultrusion, considering their use in a highly efficient, continuous production process for composite goods with a complex cross-sectional profile. It is shown that very few calculation methods exist for determination of the technological parameters or the required structural characteristics of such materials. Proceeding from the analysis performed, it is proposed that the unified braiding pultrusion technological process be used with broaching methods to increase the means of manufacturing complex-configuration goods.

Chapter 2 is devoted to investigations in kinematic characteristics of the braiding process of composite layers. It is emphasized that the distinctive feature of braided layer formation is the effect of geometric parameters (number of plaits, braiding angle, article diameter) on the plait section shape in a layer and on the filling degree. A calculation method and relations for the filling degree computation, limiting the filling degree values for the layers with the maximum surface density, are described. It is also noted that the theoretical results are consistent with experimental investigations in the filling degree values under different longitudinal-to-braiding layer ratios. These form the basis for recommending the method developed for defining structural parameters of the materials.

Chapter 3 presents the analysis of choosing the efficient pultrusion process parameters to form complex-reinforced rods. Probable variation ranges of the time-temperature regime during molding are shown, and major requirements for the components used are formulated. Investigation results of temperature fields in the spinneret working channel and their regulation with the help of heating units are described.

Process parameters were chosen on the basis of experimental results of the binder setting in response to drawing velocity and spinneret temperature. Based on the derived data, most of the optimum drawing velocities were chosen from the viewpoint of the highest setting degree.

A calculation model of the stress–strain state of the profiles is proposed in Chap. 4. By using the model, it is possible to explain failure mechanisms and to calculate structural elements of variable thickness. It is reported that tangential stresses in their concentration beds along with normal stresses in their maximum points are highly dangerous for the articles. It is emphasized that the tangential stress value is half that in articles formed by braiding pultrusion than in those without a closed braided layer. Mechanical fixing of braided layers by broaching eliminates the stress concentration in the critical zone and improves the strength characteristics of the profile under compressive loads. The main dimensional and structural parameters of standard integral complex-profile rod designs have been determined, and a calculation method with allowances for the effects of technological and service factors is set forth in Chap. 4.

Chapter 5 cites the results of studying the resistance force to material movement versus major technological parameters of the process. On the basis of force factor analysis, it was determined that the total resistance is composed of the resistance forces on separate portions of the technological line where, apart from frictional forces, movement of the material is hampered by the forces of its cohesion with the mandrel. Separate elements of the technological process are studied, and a problem for resistance in the spinneret simulating article molding is solved, taking into account the volume of shrinkage of the binder and the change of the article's geometric parameters during thermal treatment.

In Chap. 6 are presented calculation methods for the force parameters of the molding process of a complex-profile article. It is proved that molding of the complex-profile rod-like articles is feasible, providing that separation of the materials of the profile elements is observed. It is suggested that the materials be separated using special variable-width plates with simultaneous broaching. This improves mechanical characteristics of the article and ensures high-quality final operation of flange molding. Based on the analytical results and experimental verification, recommendations on the design of the technological rigging and equipment were worked out.

The main characteristics of broaching operation of composite materials are analyzed in Chap. 7. On the basis of the analysis, a single-lane method of broaching has been selected as the chief method for production of complex profiles. Estimation of the adhesive seam strength has shown that a simplified scheme of the joint can be used without mechanical fixing of stitches. For this purpose, simpler equipment can be used to realize the required strength of the joint. A method developed for estimation of the seam joint strength has shown that strength and design parameters are interrelated. A special calculation procedure has been elaborated to estimate the forces and pressures that arise in the needlebroaching material pair, which is also used to find pretension of broaching threads or plaits while taking into account the dynamic and kinematic factors of the process.

Chapter 8 is devoted to elaboration of the route technology of manufacturing complex-profile composite goods by braiding pultrusion molding. Recommendations are presented for choosing special devices that monitor the process continuity and for realization of preset technological and design parameters. Methods for computing the geometric parameters of the rigging and equipment are proposed. Test results of the technological process as well as comparative experimental investigations have proven that the method and material structure developed significantly increase the mechanical characteristics of the complex-profile articles under various loading types.

Investigation results of the physicomechanical characteristics of the composites, depending on their structural and geometric characteristics, as well as technological regimes are reported in Chap. 9. The results obtained are in full agreement with microstructural investigations. The stress-strain problem of an article during broaching has been solved to improve the performance of the broaching mechanism. A general methodology for the choice of the main technological parameters of the braiding pultrusion process is set forth. The methodology makes allowance for limitations imposed on the structural and geometric parameters of the material, the kinematic characteristics of the equipment, the optimum features of thermal treatment at different stages, and the force parameters of the molding.

In Chap. 10, application of the technological process developed for rod structures in the mining industry is described. Taking into consideration the operation conditions of dead-end rods for excavations, it is suggested that spirally reinforced fillers be used for their manufacture. Thus far, it is necessary that development of a technological process for such materials ensures the production of zero-defect layers of the longitudinal spirally reinforced filler. In the course of investigations, molding regimes of the rods were determined and their physicomechanical characteristics under long-term loading were studied. An anchor design is proposed that takes into account roof installation and bolting.

Chapter

1

Production of Composite Complex-Reinforced Structures by Continuous Methods

Various rod-like elements are frequently used in structures with different purposes. Examples are aircraft parts, including the local reinforcements made of high-modular, high-strength materials used as stiffening elements in most loaded areas.[76,423] Reinforcing elements can be installed by traditional methods that often help to cut use of costly composite materials by employing glued joints.[4,306] Airframe designs commonly employ stiffened casings of composite materials barely half the weight of aluminum. The reinforcing elements for stringers and frames are mainly π, T, and Z profiles, as well as box-shaped or unbroken sections. In addition to profiles made of a single material grade, combined structures based on metal composites, only metals, or only composites are also available.

Stringers and frames made of unidirectional materials were found to impart high specific strength to their structures,[140] although they were at risk of axial splitting.[81,212,229,253,359,397] To this end, orthogonally reinforced materials were used with high transversal stiffness. The production technology of such stringers involves preliminary molding of a laminate plate consisting of layers of different materials, which can be unidirectional layers of carbon fibers and layers of glass fabric. The plate is then cut into strips of the stringer cross-sectional size. This process is labor-intensive, and the product is not always of good quality. Of specific interest is production of combined stringer and frame structures based on boron fibers.[243] These highly modular boron fibers are

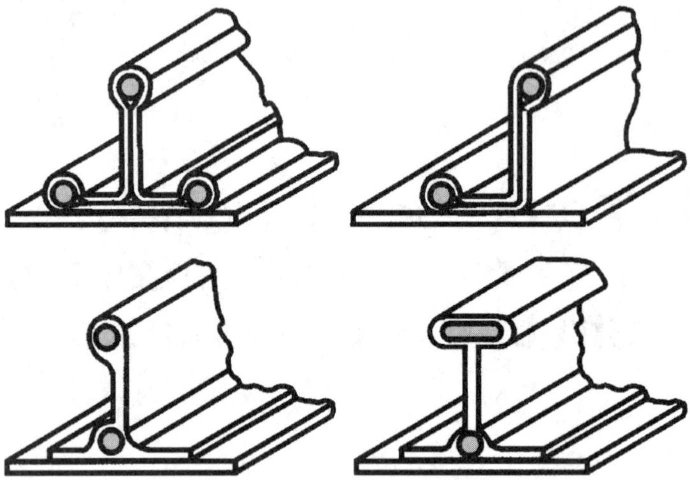

Figure 1.1 Reinforcing elements of force structures.

placed precisely in an element cross section to improve its strength and stiffness. Named products can be manufactured by pultrusion, bending, molding, or assembly of shaping profiles. Sometimes, the weight of the combined stringer structures is 2.5 to 3 times lighter than the aluminum designs. However, a number of complex problems arise during production of such stiffening elements. Some are associated with residual stresses, leading to distortion of the part shape, while others are associated with the choice of the binder, which should be at the lowest possible temperature. Some modifications of named products are illustrated in Fig. 1.1.

In addition, a series of requirements with respect to strength, stiffness, heat resistance, and so on are imposed on these types of articles,[35,93,387] which can be met by continuous highly productive processes.[76,106,370] The problem of developing continuous efficient methods to mold goods from composite materials with complex structural reinforcement schemes has recently come to the forefront. Analysis of literature has shown that braiding[60,295,399,400,426] and pultrusion[5,30,31,42,69,164,198,240] are among the most critical points. Certain restrictions are put on both continuous methods as well. Although braiding creates a spatially reinforced skeleton of the material, it is difficult to produce the required length of articles or to gain necessary productivity.[266,334,394] On the other hand, pultrusion presents a continuous molding process, but serious difficulties are encountered at creation of a complex reinforced material structure.[90,233,252,293] It is apparent that the necessary effect can be achieved if the two processes are combined,[370] where the continuity and quality of articles will be guaranteed by pultrusion. As for braiding, it creates the conditions for the pultrusion of articles with a given structure from the one hand while providing for a given reinforcement scheme on the other. It should be noted that attraction of these two methods is sometimes insufficient for achieving certain structures. In this connection, some additional technological means are used to form an article with needed service characteristics. For example, to produce complex-profile products like channels or flange beams, one should envisage the stages of fixing separate elements of the profile to avoid peeling under loading. Joining the two methods into a unified technological process was not employed in practice for production of rod-like articles until the present time. Let us consider here the indicated forming processes to examine their peculiarities and the potential of their combination in a single technological process.

1.1 Service Characteristics of Composite Profile Articles

The structure of fiber-reinforced materials, i.e., fibrous composites, represents a system in which one component with high mechanical strength and stiffness transmits the force flows in the construction. Another component is a binder that adds continuity to the system, joins separate fibers together, and transmits strains from fiber to fiber in transverse directions, making grounds for their simultaneous deformation under loading.[51,139,244,396] Properties of reinforced materials having such structures can be varied widely and can be strengthened in most loaded points to meet operation demands.[1,2,86,258,421]

The glass fiber fillers were the first to be used in industry as material reinforcements. In recent years, composites based on boron, carbon, organic materials, and other types of fibrous fillers have come into intensive use. Numerous base and supplementary parts are produced from these materials today.[244]

Consumption of composite materials in various domains is rising steadily. The yearly increase in application of promising composites is said to be 20 to 25 percent. A couple obvious reasons are breakthroughs in fuel economy and depreciation of reinforcing fillers production. The cost of carbon and organic fibrous fillers has come down to a level that makes them promising and expedient for adaptation in aviation and rocket production as well as in engineering, shipbuilding, medicine, construction, and so on.[76,94,137]

Profile articles prevail as elements of different designs. They are employed both as individual parts, e.g., beams, shafts, and rods, and as reinforcing elements, such as stringers, longerons, frames, and so on, thanks to their insulating or conducting properties in combination with their high mechanical qualities. The large-scale consumers of profile articles include transportation, collection, and water-supply systems, naval fleets, construction, mining, and electrical power and electrical industries.

The type of a profile is conditioned by a definite structure of an article and its operation conditions. Cross sections of rod-like elements can form solid, hollow, closed, or open profiles, and they can be simple or complex-shaped geometrical figures (Fig. 1.2). The decorative and low-loaded profiles are not bound by very strict requirements with respect to their mechanical properties.

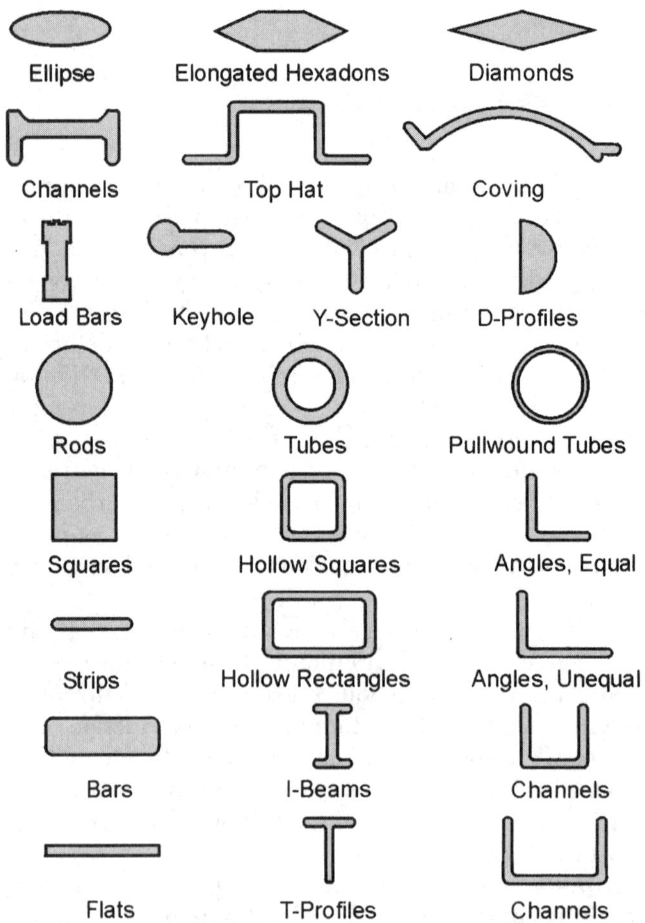

Figure 1.2 Main types of profile ware based on composites.

Most optimum structures used at simple loading modes (tension, compression, torsion) are rod elements with either round or annular cross section. They are perfectly balanced mechanically and thermally, and they are easy to manufacture.

Structures operating under bending and combined loading experience most complex types of mechanical effects, and their cross sections experience both normal and tangential stresses. Because the stresses are spread unevenly over their sectional area,[38,41,45,85,153,328,355] the optimum structure should acquire a complex profile shape. Frequently employed for operation under bending are double-tee, channel, box-shaped, and Z-shaped in cross-sectional profiles. When rods are used as stiffening elements, they are most often V- and T-shaped profiles with a flange fit for fastening to a supported plate or cover.[258,346]

During application of bending loads, all types of profiles experience maximum tensile and compressive axial stresses largely perceived by the profile flanges in the points most remote from the neutral line (Fig. 1.3). Axial stresses generated in the profile post reach their maxima in points on the neutral axis. Horizontal tangential stresses also appear in the flanges of thin-walled complex-profiled articles and change by the linear law, reaching their maxima where the flanges abut the post.

In isotropic materials, tangential stresses are rather low compared to normal ones, and they can be neglected. When fiber-reinforced composites are used in profiles, their shear strength is commensurable to the acting tangential stresses, and the bearing capacity of such articles is significantly reduced.[122,166,184,350] This is mostly related to materials

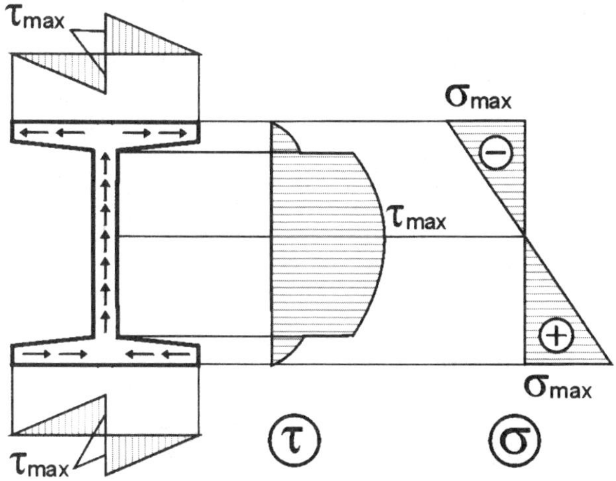

Figure 1.3 Distribution of normal and tangential stresses in complex-profile rod elements.

based on high-modular and high-strength fillers, including organoplastics, boron plastics, and carbon plastics whose anisotropy reaches 40 to 100.[73,144,391] This is because mechanical characteristics of reinforced materials are conditioned at shear in the transverse direction chiefly by the mechanical properties of the polymer matrix and the cohesive strength of components, which is much less that of the fibrous fillers.[47,227,245,321,352,391,392,418,420]

Elevation of transverse and shear characteristics of reinforced polymer materials is a current trend of investigations in the field of manufacturing composite goods.[13,62,142,246,321] The named characteristics are critical at compression and bending of composite materials.[8,53,278,284,348,421] Interlayer shear-induced failure is often observed in such structures as rods, plates, shells, etc. because they lose stability in the zones of edge effects under a point load application[41,100,341,348,402,412,429] or thermal effects.[49,259]

Poor resistance to transverse separation is the reason behind delamination of structures under comparatively light transverse stresses.[45,79,137,144,162,169,274,356,362,363,364] This may lead to bulging of individual layers under compression or bending,[40,43,141,146,305,360,361,409,410] as it is with e.g., wound structures under the outer pressure effect.

Along with the creation of new types of reinforcements and important binders to perfect the mechanical characteristics of reinforced plastics, there has been an improvement of the design schemes of composite products.[188,258,264,363] With this aim, it is necessary to take into account the production technology, be it braiding,[187,264,286,309,335,406,428] textile manufacture,[20,61,188,221,377] knitting, [286] or pultrusion,[32,88,89,90,101,215] to develop novel technological processes making maximum use of the initial components' strength.[363,385] Engineering procedures aimed at improving the characteristics of composites presuppose introduction of an additional filler at an angle to the main reinforcement. This results in improved mechanical characteristics in transverse directions and additional bonding between the main reinforcing material layers. Investigations in this field have taken the paths of creating layered[38,39,226,365,405] and multidirectional structures,[84,190,324,404,413,417] braided and spatial-reinforced materials,[103,105,147,343,399] and using of hybrid composites[193,294,390] with the purpose of attaining high mechanical characteristics of the composites in directions of the main stresses.[108,109,111,112,114,115,116,185,348]

Along with a number of advantages, the methods discussed herein bring about a set of negative aftereffects. Introduction of an additional material in the same direction as the main reinforcement results in impairment of the composite mechanical characteristics in this direction.[22,113,145,291,389] Besides, production of spatially reinforced and woven structures leads to flexure of the fibers, which worsens the mechanical characteristics.[159,226,269,378]

1.2 Braiding as a Means of Creating Spatially Reinforced Material Designs

In a general case, it is evident that to attain the required properties of a material a spatial reinforcement structure must be formed. Such a structure can ensure certain characteristics of the material in preset directions, which is supported by numerous investigations for different materials,[23,188,244,348] loading conditions,[187,387,403] and production procedures.[56,140,242,254,256,265,343,395]

Classification of spatially reinforced plastics is proposed in Ref. 348, in which the materials formed by a system of two or three threads and whiskered reinforcing fibers are marked out (Fig. 1.4).

Braided materials (a system of two threads) improve strength and stiffness of ready products at interlayer shearing by 2 to 2.5 times.[148,151,169,178,280,318] One drawback of this method is flexure of the reinforcement, which worsens properties in the basic direction with thickness increment. Thick-walled articles can be produced by forming a reinforcing skeleton from a system of three threads (Fig. 1.5).

Manufacture of products based on assembled rod carcasses has been described elsewhere.[348] For the large-size articles, a carcass based on 3, 4, or 11 rectangular threads is used.[82,171,348] In the case with

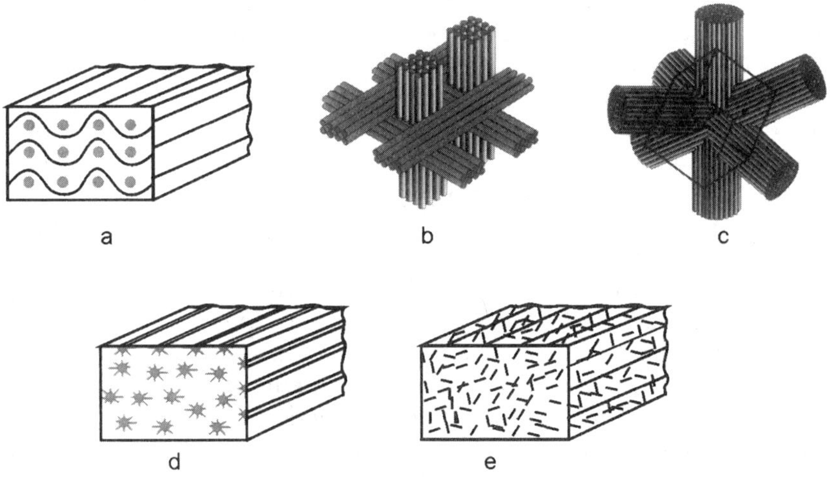

Figure 1.4 Major procedures of spatial reinforcement: (*a*) in two threads; (*b*) in three threads; (*c*) in four threads; (*d*) application of whiskered fibers; and (*e*) use of filamentary crystals.

Figure 1.5 Nontraditional construction: a 3D braided system.

three rectangular threads, special equipment is used to perform three interrelated motions, or the carcass is braided in a plane around tubes or dowels, followed by dragging the reinforcement through in the third direction.[348] When tubes are employed, the material is damaged less than when dowels are used; however, in both cases the material lying in the other two directions is damaged. In the latter case, when 4 or 11 rectangular threads are used, the carcass is assembled by hand using special devices.

In all cases mentioned, elastic and strength characteristics of the material in transverse directions improve significantly. According to data presented,[165,168,192,213,320,342,343,419] the strength of the materials reinforced by a system of three threads improves in transverse directions by four to five times, the interlayer shear modulus improves by one to two times, and the strength at interlayer shear increases 2 to 2.5 times. Analogous results were obtained in Ref. 286, where a knitted fabric was used to increase the strength in transverse directions. It was discovered that the strength and interlayer shear modulus of the material improved by 45 and 15 percent, respectively, even with few fibers in the transverse direction.

The composites reinforced by whiskered fibers[288] create interlayer bonds via filamentary crystals grown on the fiber surface or impregnated into the binder. The mechanical properties of such composites depend largely on the type and amount of the filamentary crystals, binder composition, type and volume content of the fibers, and on compacting pressure. For instance, when up to 5.2 wt% of filamentary crystals of SiC is impregnated into a carbon plastic, its interlayer shear strength rises from 16 to 82 MPa under an intact elasticity modulus in the basic direction.

The computation procedure of spatially reinforced materials is reported for knitted structures,[128,161,163,304,386,401] for woven structures,[134,168,220,223,292,313] and for braided structures.[21,173,213,273,342,422,423] The computation results agree perfectly with experimental data for different materials. Deformation characteristics of the materials formed by systems of two and three threads have been estimated.[57,59,130,163,338,375,378,380,381] A method of forecasting efficient constants of composite materials reinforced by spatially oriented short fibers is described in Ref. 381.

From the procedures described above, it follows that, in order to improve the transverse characteristics of materials, it is worthwhile to use 3D reinforcement schemes. Note that various reinforcement schemes are acceptable for the thin-walled structures, e.g., oblique-angled reinforcement, reinforcement with filamentary crystals, or with a system of two or three threads.[353] For thick-walled structures, application of whiskered fibers or reinforcement with a system of two or three threads is recommended. In all cases mentioned, definite improvements in characteristics of the materials and products are attained, although each of the methods touched upon has its difficulties in implementation. Apart from this, a number of procedures have been developed recently based on the 3D braiding technique.[3,123,197,281,419]

Spatial braiding is a production technique for the 3D reinforced structures obtained by a continuous braiding of fibers. All fiber carriers move simultaneously during the braiding process, which is a distinctive feature of braiding in contrast to weaving, where all threads of the base are kept by an immovable frame and each fiber is threaded by a shuttle. Numerous carriers of the reinforcing fibrous material are employed during spatial braiding (Fig. 1.6).

During motion over the base plate guides, the threads or plaits are interweaving. The guides are installed to comply with braiding either in cylindrical or in Cartesian coordinates. The mechanisms of spatial braiding have been set forth.[75,92,189] It is stated, in particular, that when the device is used for braiding in cylindrical coordinates, braided products can be manufactured in the form of large-diameter long-length cylinders. This device contains 15 concentric circular guides having 144 radial slits in which 2016 fiber carriers are located. The braiding device in Cartesian coordinates incorporates 64 rectilinear guides with 19 slits having 6224 fiber carriers. When a 3D braided carcass is employed, a quasi-isotropic material can be created in which delamination planes are absent. It is quite evident, judging from the number of fiber carriers only, that these devices are rather complex for production, maintenance, and operation.

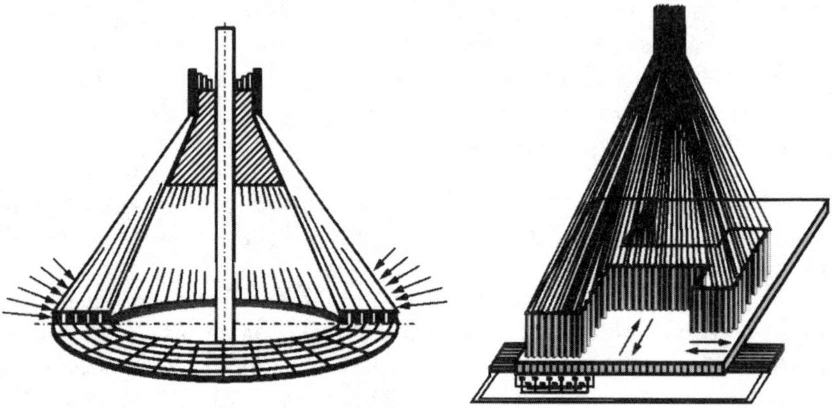

Figure 1.6 3D braiding devices: (*a*) braiding in cylindrical coordinates; (*b*) braiding in Cartesian coordinates.

One example of a comparatively simple automatic braiding technique for articles with a 3D reinforcing carcass is the one reported in Ref. 64. This procedure is in use at Morton Thiokol Company for production of heatproof pipes.

The carcass is braided from a carbonic felt on a mandrel whose diameter is similar to the article's inner diameter. Rods of carbon fibers are placed on the mandrel surface perpendicularly to its axis and evenly arranged over the needed part of the mandrel. These rods serve in the ready product as the transverse reinforcement. The prepared mandrel is then braided with layers of carbon plaits. Each layer contains interlaced plaits of three directions (3D texture), namely braiding plaits (left and right) and axial plaits. The braided layer acquires a cellular structure in which each cell is formed as a result of interweaving of the plaits of the right and left directions whereas the axial plaits are located over diagonals of the cells. The process terminates as soon as the needed number of layers is applied, at which time the article is heat-treated and machined (Fig. 1.7).

According to Ref. 64, the braider used for production of block products is based on a standard machine employed for such braided articles as cables, high-pressure hoses, and other items. A modernized braider contains three groups of fibrous reinforcing material carriers: the first one is a group of immovable carriers and is intended for the axial reinforcement, while the other two groups carry the braiding reinforcement. The movable carriers with the braiding reinforcement are placed over circular undulatory trajectories, one group clockwise, the other group counterclockwise. The braider described in Ref. 64 contains 144 carriers of fibrous material, 72 of which are stationary.

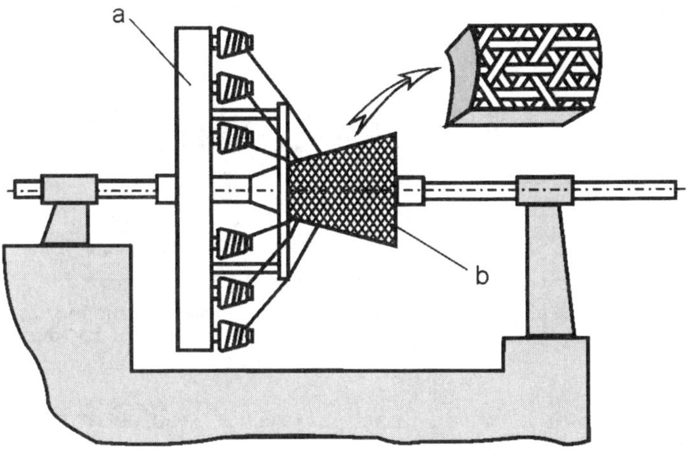

Figure 1.7 Braiding over the mandrel: (*a*) braiding unit; (*b*) mandrel of carbonic felt with rods.

Among undisputable merits of the method described above is its level of automation. Nevertheless, braiding on a mandrel with preliminary arranged rods has some drawbacks. The surface shaped by the plaits quitting the braider can cross over with the rods and lead to mutual damage of the reinforcing materials.

This imposes limitations on variations of reinforcement parameters, such as reinforcement angles, radial rod pitch, and so on, and toughens the requirements for the kinematic stability of the process. As a result, the pitch of the reinforcement structure of the rods is preset and cannot be changed in subsequent layers. Moreover, use of this method in continuous conditions is highly problematic.

Analysis of literature has shown that, when producing rod-based articles with specific structures, one can choose a number of available automatic machines that can braid the reinforcing filler. In the majority of publications, thermal treatment is suggested after winding in a heat chamber followed by machining. In principle, the equipment considered renders braiding for only discrete products, though after certain development, the equipment can also be used for continuous braiding.

1.3 Pultrusion for Continuous Molding of Composite Ware

Pultrusion of fibrous polymer composites first appeared in 1948 as applied to the glass-fiber-reinforced plastics on the base of setting binders.[94,164,224,240,241,252,302] It was modified in the early 1960s when the growing demand for profile products of polymer composite materials encouraged development of high-speed production methods.[46,80,233,328,339,425]

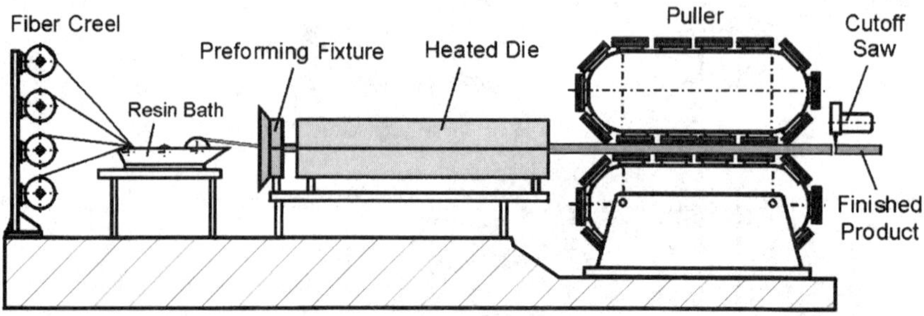

Figure 1.8 Functional diagram of pultrusion process: (1) reels with fibers; (2) impregnating bath; (3) preforming dies; (4) heated forming die; (5) puller; and (6) finished product.

A functional diagram of the pultrusion process is illustrated in Fig. 1.8, the main stages of which are:[164,203]

Unreeling of the fibrous filler from the bobbin carrier

Impregnation of reinforcing fibers with a polymer binder

Molding of a blank in preforming dies

Final molding of the profile and shape fixing via setting the polymer matrix in a heated forming die

Profile pulling and cutting into needed length pieces

Additional thermal treatment of the articles (in case of necessity)

The reinforcing fibrous filler is pulled from the bobbin carrier at a given speed into the bath with a binder and the preforming zone where the fibers are formed into the needed profile shape, compacted, and the excess binder is removed. Afterward, the material is then fed into a metal die heated to the required temperature to ensure a given rate and depth of the binder setting. The profile is finally formed, and its shape is fixed, whereupon the long-length article is fed by the puller into the cutter. Profiles obtained in this way correspond precisely to the required cross section and do not require any further machining.

Several types of pultrusion machines known in the art are distinguished by the following features: [330,366,367,368,369,370]

Periodicity and continuity of pulling

Horizontal or vertical type of installation

Type of puller

Peculiarities of molding and setting zones (presence of a winder, coiler, type of die, and its heating method)

Impregnation method of the fibrous reinforcement

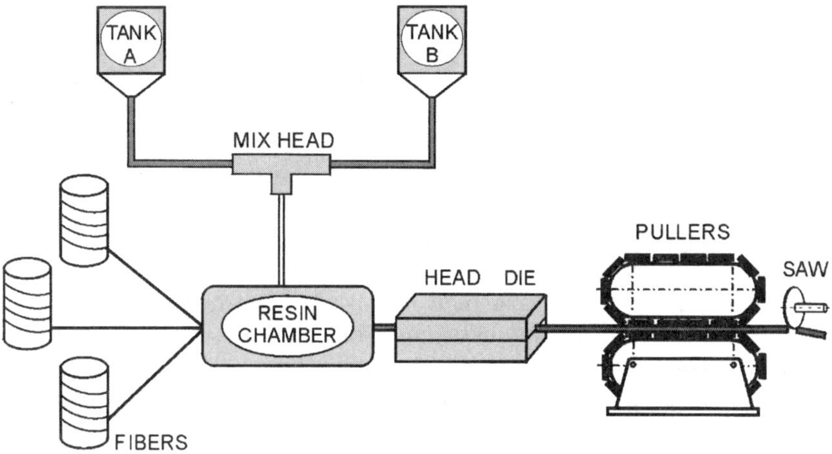

Figure 1.9 Reaction-injection pultrusion.

The majority of pultrusion machines are horizontal. Vertical apparatuses have been designed for the manufacture of profiles with a solid cross section and a hollow core.

The bobbin carrier is intended for the reels that carry fibers and is made as a metal frame with horizontal axes. To install large reels with glass fibers, for example, block-type bobbin carriers are used. In this case, the fibers are unreeled inside the reels.

The fibrous reinforcement is usually impregnated[83,207] when it is pulled through the impregnation bath, which contains a polymer binder and a system of pinch and squeeze rollers. To lower the viscosity of the binder and improve fiber wetting, the bath is equipped with heaters. To refine the impregnation process, a number of injection methods of the binder have been developed,[207] one of which is presented in Fig. 1.9.

The forming die represents a built-up mold whose working surfaces have been polished and chrome-plated. The die length, which is in line with the profile pulled, binder type, and process speed, is about 300 to 1500 mm. The working channel has a regular geometrical shape with parallel walls and rounded-off edges to avoid fiber rupture.[119,164,317] The die is commonly heated by flat electric heaters. To increase the efficiency and pultrusion rate, high-frequency induction heating is being employed ever more often in industry.[48,66,132,241,411] The most important parameters of die heating are uniformity of temperature distribution and temperature control at all molding stages.[67,132,214,303,372,373]

The pullers are based on the simplest forward thrust exercised by winding a rope or chain onto a drum. Two types of puller are used today, and they are different in principle. In the first type, two caterpillar tracks are continuously revolving in opposite directions and a profile is placed between them. One apparent drawback of this mechanism is the necessity of a great number of specially formed clamping pads for the profiles. The second reciprocating device seems to be more convenient as only two pairs of pads, which correspond to each profile kind, are installed in the clamping system. One pad is pressed against another by a hydraulic or mechanical drive, and the profile is found between them.[29,80]

Pultrusion machines are used in the electrical industry chiefly for manufacture of round, glass-reinforced plastic rods of 10 to 50 mm diameter. The most well-known Russian machine of this type is the 4-UPS-12, which includes two benches for bobbins, chambers for plait drying, impregnation, and molding, a cutting device, a reception facility, and a polymerization chamber. Its throughput rate is 0.3 to 0.6 mm/min at a forming die length of 2.4 mm.[310] Other pultrusion installations have also been described.[29,199,207,307,328,339,340,374]

Pultrusion processes are presently under intensive development in the United States, where more than 150 production lines are functioning today. The leading companies in this domain are Goldsworthy Engineering and Pultrusion Technology, which are manufacturing standard machines to achieve up to 150 cm^2 in cross-sectional profiles. Goldsworthy Engineering has devised one of the largest-scale installations for production of profile panels from hybrid materials of 457- by 914-mm section and up to 10 tons of pulling force. The installation is equipped with a high-quality induction heating system to solidify the profile material.

For production of tubular or rod-like elements, pultrusion and winding with a thread are combined to improve shear strength and resistance to torsion. As shown in Fig. 1.10, rather intricate equipment is necessary for this method, and the processes themselves must be comprehensively studied.

An example of a pultrusion machine for manufacture of long-length profiles developed by Faserplastik Company is presented in Fig. 1.11.

The machine of the company "Faserplastik" operates as follows. The threads from bobbin carrier (1) are fed through grillage (2), separator rolls (3), and pilot holes (4), such that they will lie parallel when they enter the impregnation bath (6).

To improve the impregnation quality of the fibrous filler by the polymer binder, the filler and binder are dried by an IR lamp located above the fibers bundle before the inlet to the impregnation bath. The distance from the fiber bunch to the lamp is regulated to make the convection heat current carried by the filler stabilize the heating of the binder through the impregnation bath at the needed temperature.

After the bath, the fiber bunch enters the system of preforming dies (7) and director (9), where the filler is further saturated and excess of the

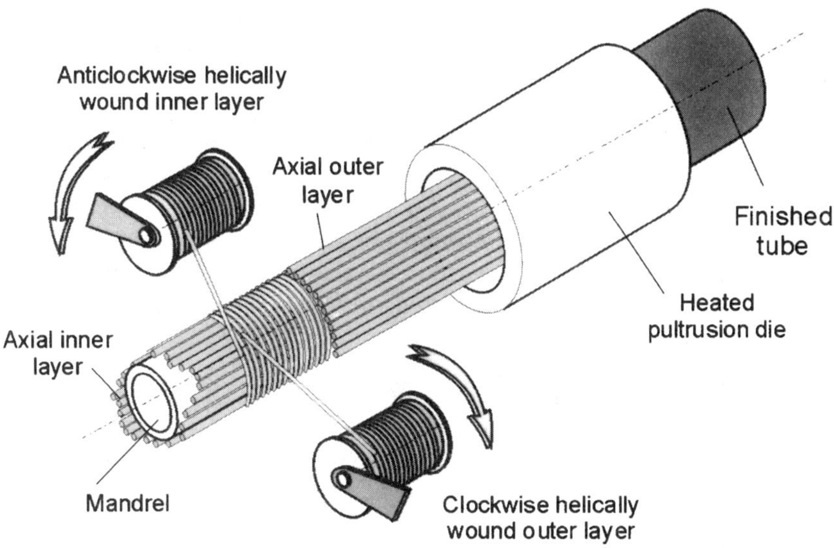

Figure 1.10 Pull winding process.

binder and bubbles are removed. Individual fibers are also aligned here and distributed across the section, and the profile section is molded. In addition, in the bunch of fibers with a slight excess of the binder, which is fed to a heated die where the remainders of binder and air are removed fully, the fibers are aligned over the profile section and the binder is set. The die represents a 450-mm-long metal detachable mold with channels of the required cross section. The described machine is designated for manufacture of unidirectional articles only.

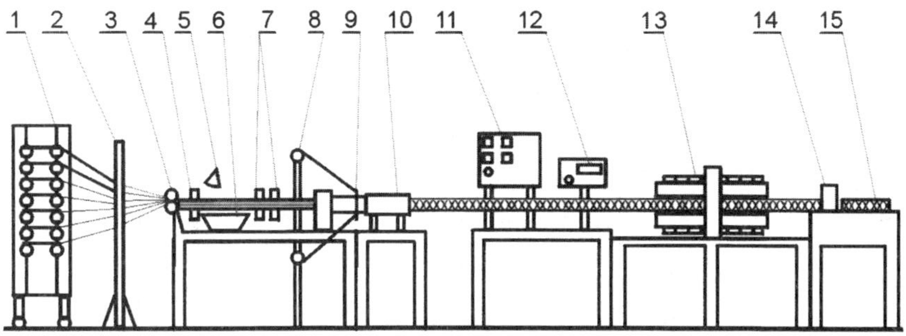

Figure 1.11 Functional diagram of Faserplastik pultrusion machine: (1) bobbin carrier; (2) grillage; (3) separator rolls; (4) pilot holes; (5) IR lamp; (6) impregnation bath; (7) preforming dies; (8) reels with reinforcing strips; (9) director; (10) forming die with a mandrel; (11) thermoregulator; (12) control panel; (13) caterpillar traction; (14) cutter; and (15) ready profile.

Many published works are devoted to investigations of the mechanical characteristics and their interrelation with other process parameters for the described products.[30,33,42,89,101,204,314,362]

Higher-quality products are produced in rather long (about 2-m) thermal chamber dies. Stable geometric parameters of such products are attained when binders with low volatile matter content are used. This procedure has not, however, gained popularity because of its costly production and the complex maintenance of such dies. Moreover, binders containing solvents yield porous and unstable articles, as it is difficult to remove solvent vapors from the closed volume of the thermal chambers during thermal treatment.

Similar procedures for forming the described articles from composites presuppose pultrusion, during which the cross section and filling degree of an article by a fibrous reinforcement are dependent upon the molding force. Most applicable is the method in which the reinforcing filler is spread over an adhesive tape whose width corresponds to the article's cross-sectional perimeter. The tape is set in motion by a special pulling device and turns up, forming a shell that envelops the reinforcing material. The material is molded either by a forming die (Fig. 1.12) or by an auger extruder at a growing reduction strain (Fig. 1.13) to remove gaseous inclusions from the structure. Primarily articles with a large cross-sectional area are manufactured by this method. An antiadhesive film also used in this procedure assists in smooth motion of the article through the die.

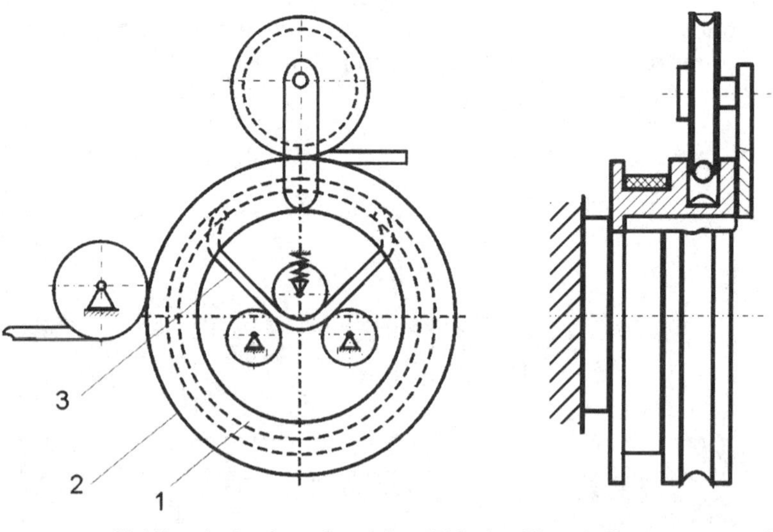

Figure 1.12 Molding device for rod articles: (1) body; (2) antiadhesive tape; and (3) pull-in rollers.

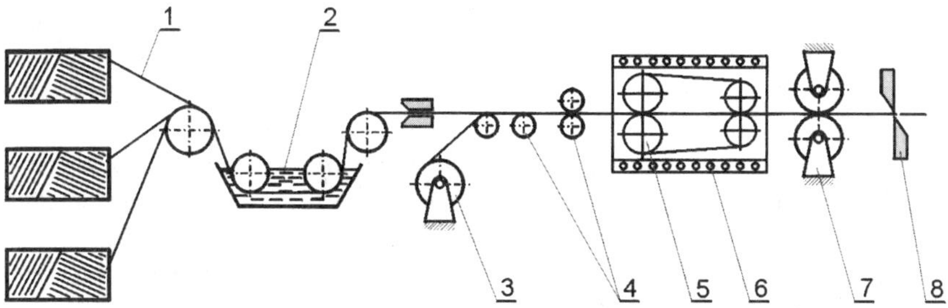

Figure 1.13 Molding of rod articles using auger extruder: (1) fibrous material; (2) binder bath; (3) bobbin with antiadhesive tape; (4) molding-on rollers; (5) auger extruder; (6) thermal chamber; (7) stretching rollers; and (8) cutter.

One pultrusion method includes winding of a fibrous material at the thermal treatment. The degree of compaction reached in the dies is fixed by the tension of the auxiliary reinforcement fibers. The highest filling degree is attained when an article is pulled through a cascade of discrete short dies with gradually changing diameter of the forming hole and is continuously heated. The functional diagram of the process is shown in Fig. 1.14. A similar method is used to manufacture high-strength glass-reinforced plastic rods intended for reinforcing glass, plastic, concrete, and electric insulating structures rather than using metal reinforcement and to manufacture plastic rods with decorative patterns.

The pultrusion methods described herein can solve the problem of continuous production of rods with simple cross sections and reinforcement schemes. Manufacture of complex-reinforced structures can significantly complicate the technology and equipment used. The resultant products show low compression and shear characteristics and, consequently, are limited in use.

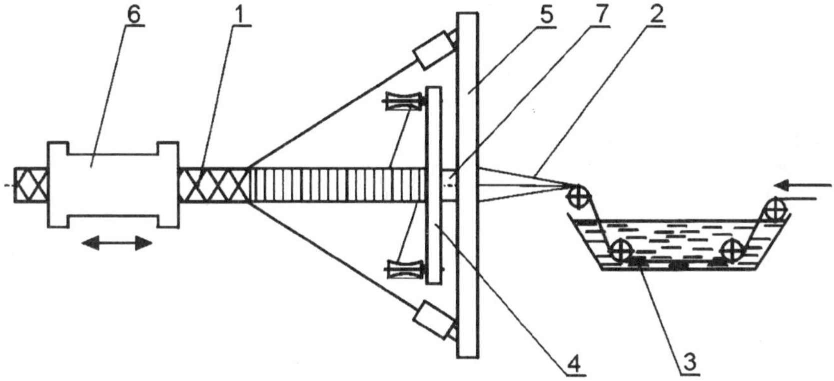

Figure 1.14 Pultrusion of composite rods with winding: (1) winding reinforcement; (2) fibrous material; (3) binder bath; (4, 5) winders; (6) puller; and (7) mandrel.

Attempts to complicate the reinforcement scheme by using tapes, mats, or reinforcement laid at a small angle to the article axis are making, in some cases, the pulling process improbable. Intricate, e.g., spatial, constructions involve numerous problems and, in fact, such constructions have not been considered for continuous processes. Nevertheless, these types of products are of greater practical interest concerning the advantages of composite materials and their feasibility in reinforcing various products in directions that coincide with loading. This problem can be solved by uniting the two highly productive automated processes of braiding and pultrusion. Production of a braided layer on the surface of an article, inside of which a necessary scheme of reinforcement has been implemented, allows the article that is being molded to move through the die under definite conditions. This layer within a material structure produced in this way furnishes additional advantages for optimizing its properties.

Analyses of braiding and pultrusion processes have formed the basis for applying the information obtained in research in estimating the molding regimes at each stage. This is, however, insufficient for designing a combined technological process that takes into account the structural and geometric characteristics of the product and their interrelation with the technological parameters. However, it remains necessary to define the limitations for combined processes, as with any other method.

The review above proves once again that, although technological methods of braiding and pultrusion exist in practice, they are able to produce only a limited range of composite articles. It should be emphasized as well that braiding has been described primarily for short shell structures without any theoretical substantiation of the kinematic potentialities with respect to continuous molding. As for the existing pultrusion methods, they are aimed instead at manufacturing long-length articles with simple reinforcement schemes, which presupposes unidirectional packing of the reinforcement in limited amounts to improve transversal characteristics. Each of the named processes yields articles with a limited range of physicochemical characteristics and production volumes. The present authors suggest, as indicated earlier, that the three procedures of braiding, pultrusion, and insertion be combined into a single technological process. Based on investigations already carried out, the problems of developing the technological processes for production of composite material rod structural elements by a combined braiding pultrusion method are considered in this book.

Chapter

2

Investigation of the Braiding Process of Composite Layers

The geometric characteristics of the structures of braided composites have been studied comprehensively.[187,343,398,400] Apart from braided structures, 3D designs,[188,399] knitted materials,[286,91] and semis based on textile materials[56,129,135,222,378,379] have been investigated at length. These cited works have treated primarily the architecture of the fibrous material distribution in a semi and the manufacturability of the material during the process.[217,243,329] Braiding technology itself is scarcely reflected in literature. The peculiarities of manufacturing braided composites to attain desired technological and service characteristics of ready articles are discussed in this chapter.

2.1 Analysis of the Braiding Process Kinematics

Let us consider the forming conditions of the braided layer with preset structural parameters. As long as the surface layer experiences high shearing loads during pultrusion, it is expedient to form such a layer based on 2D and 3D textures, whose general views are shown in Fig. 2.1.

Most appropriate from the viewpoint of the pultrusion process is the triaxial 2D structure (Fig. 2.1*b*). In this case, along with the reinforcement laid at an angle $\pm\varphi$ to the article axis, a longitudinal reinforcement is also introduced into the layer structure to allow for variations in conditions of the article as it is pulled through the spinneret. To impart high and stable characteristics along the length of the material, it should be

a b c

Figure 2.1 Braided textures: (*a*) 2D braid; (*b*) 2D triaxial braid; (*c*) 3D braid.

of low porosity and filled evenly with the reinforcing fibers. Thus, one requirement for the braided layer of the material will be obtaining the needed surface density as well as high service characteristics and stable fixing of braids on the mandrel surface.

The maximum density of the layer is not reached unless a definite interrelation is gained between kinematic parameters of the process, namely the rotational velocity of the spindles along the guides of the braider around the braiding axis and the axial travel of the item. The required dependencies of the kinematic parameters are obtained after the shape of the article surface curve along which the braids are to be laid has been determined, provided that the surface is densely packed (Fig. 2.2*a*).[353] Let us consider the general case when the article is formed by rotation of the curve $r = f(z)$ around axis Z (Fig. 2.2*b*). The braided layer is formed from rhombic cells that contain a diagonal strand (3D structure). When this layer densely covers the surface, the distance between two neighboring inclined braids in one direction (family) measured over the wide-section circle arc is equal to the axial braid width in any cross section. Because the shape of sleeving braids laid in different directions is identical but differs only in twist direction, the article surface can then be considered as being wrapped by a single family of braids. The distance between adjacent sleeving braids is equal to the axial braid width t_0.

Let the number of axial braids be n, thus the respective number of sleeving braids of one family is $n/2$. A sleeving braid width is t_b, and the distance between neighboring braids is t_0. The maximum surface density in a cross-section of r_t radius will then be of the analytical form

$$\sin \varphi_a = \frac{n \cdot t_b}{4\pi \cdot r_t - n \cdot t_0} \tag{2.1}$$

It is evident that the braids can be located on that part of an article

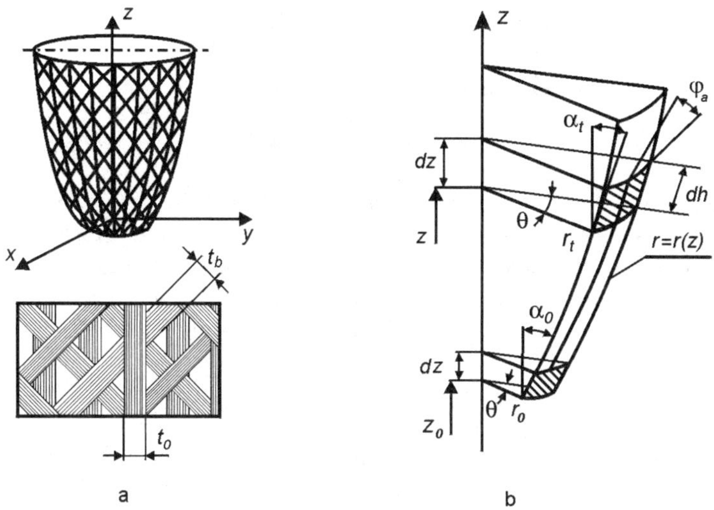

Figure 2.2 Determination of parameters of a constant surface density layer: (*a*) shell surface; (*b*) variables to the equation for determining braid shape in a constant surface density layer.

for which the following condition is met:

$$r \geq r_0 \tag{2.2}$$

where r is the circle radius of an arbitrary latitudinal cross-section and

$$r_0 = \frac{n}{4\pi}(t_b + t_0) \qquad \text{if} \quad t_0 < t_b$$

or

$$r_0 = \frac{n}{2\pi} t_0 \qquad \text{if} \quad t_0 \geq t_b$$

In r_0, all braids are directed along the generatrix, and each braid is believed to be composed of similar strips of infinitely small width that are laid evenly without distorting their shape over the surface. Consider a surface element bounded by two latitudinal planes $z = z_0$, $z = z_0 + dz$, and two neighboring meridian sections with raster angle $d\theta$ (Fig. 2.2*b*). For this surface element, the following relation will be true:

$$r_0\, d\theta \frac{dz}{\cos\alpha} = r_0\, d\theta \frac{dz}{\cos\alpha} K + r_0\, d\theta \frac{dz}{\cos\alpha}(1 - K) \tag{2.3}$$

where $\alpha =$ tangent slope angle to shell generatrix in corresponding section $\tan\alpha = r_z$

$k = t_b/(t_b + t_0)$

$r_0\,d\theta K =$ elementary strip width in section $z = z_0$

$r_0\,d\theta(1-K) =$ latitudinal distance between neighboring strips

Let us take an element on the surface restricted by two sections $z = z_t, z = z_t + dz$; and two meridian sections with raster angle $d\theta$. Because the elementary strips of the braids are evenly distributed over the surface without distortion of their initial shape, the distance between these strips is constant in any latitudinal plane. Then,

$$\sin\varphi_a = \frac{r_0 \cdot K}{r_t - r_0(1-K)} \tag{2.4}$$

where $\varphi_a =$ angle between a strip and latitudinal plane and
$r_t =$ current radius value in latitudinal section. The height of the allocated element is

$$dh = r_t\,d\theta \cdot \tan\varphi_a; \tag{2.5}$$

where

$$\tan\varphi_a = \frac{r_0 \cdot K}{\sqrt{[r_t - r_0(1-K)]^2 - r_0^2K^2}} \tag{2.6}$$

$$dh = \frac{dz}{\cos\alpha} = \frac{dz}{\cos(\arctan r_z')} \tag{2.7}$$

By substituting (2.7) and (2.6) in (2.5), we obtain

$$dz = \frac{r_t\,d\theta \cdot \cos(\arctan r_z')r_0K}{\sqrt{[r_t - r_0(1-K)]^2 - r_0^2K^2}} \tag{2.8}$$

$$\frac{d\theta}{dz} = \frac{\sqrt{[r_t - r_0(1-K)]^2 - r_0^2K^2}}{r_tr_0K \cdot \cos(\arctan r_z')} \tag{2.9}$$

The solution of differential equation (2.9) is a certain function that for a layer of 3D texture describes a right-twisting braid shape that is densely wrapping the surface. The left-twisting braids have the same shape.

Taking that, according to Eq. (2.9), $K = 1$, which corresponds to the absence of axial braids, we obtain the differential equation

$$\frac{d\theta}{dz} = \frac{\sqrt{(r_0 + z \cdot \tan\alpha)^2 - r_0}}{(r_0 + z \cdot \tan\alpha)r_0\cos\alpha} \tag{2.10}$$

The solution of the equation above is a function that describes the shape of braids in a layer of the 2D texture that densely covers the surface of the article.

2.2 Free Surface Shape of Braids

Because a pickup cannot be mounted directly on the surface of the article, some amount of free surface appears on the braids fed from the pickup of the braider onto the article mandrel. This surface (Fig. 2.3*a*) is formed by the convergent sleeving braids and the axial braids interlaced between them, where each sleeving braid is a portion of a straight-line tangent to the mandrel being braided. Certain distortions arise because of the tension of the axial braids, but they are slight and will be neglected in further calculations. Thus, a ruled surface of rotation of axial symmetry is formed relative to the *OZ* axis, which is a one-sheet hyperboloid of revolution.

Let us consider an equation for a straight line found on the hyperboloid surface. The straight line (Fig. 2.3*b*) passing through point $A(r_t, O, z_t)$ corresponds to the sleeving braid of the right family. The directing vector of this line is $\mathbf{a} = AB$. Let us determine the directing cosines of this line.

According to the scheme presented in Fig. 2.3*b*,

$$
\begin{aligned}
\cos\alpha &= \frac{\mathrm{BC}}{\mathrm{AB}} = \sin\varphi_{\mathrm{a}} \sin\alpha_{\mathrm{k}} \\
\cos\gamma &= \frac{\mathrm{AF}}{\mathrm{AB}} = \sin\varphi_{\mathrm{a}} \cos\alpha_{\mathrm{k}} \\
\cos\beta &= \cos\varphi_{\mathrm{a}}
\end{aligned}
\tag{2.11}
$$

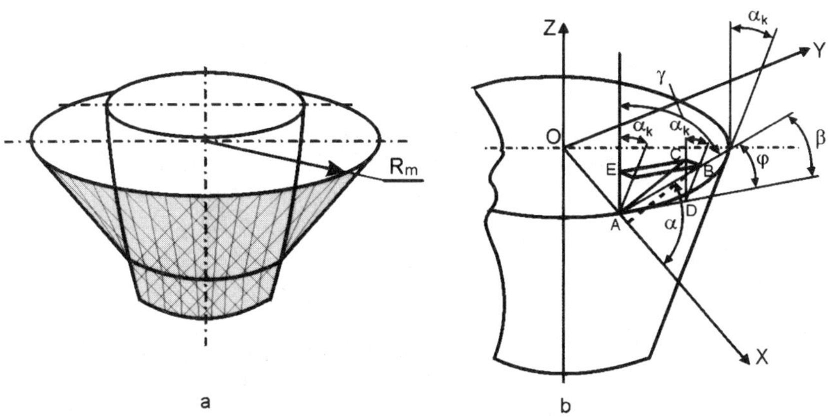

Figure 2.3 Determination of the free surface of braids: (*a*) general view of braided surface; (*b*) diagram of the equation for the free surface of braids.

where $\alpha_{\rm k}$ = slope angle of the tangent to the braided mandrel generatrix in the point considered
$\varphi_{\rm a}$ = angle between the sleeving braid direction (line *A*) and the latitudinal plane
α, β, and γ = angles between the directing vectors and *OX*, *OY*, and *OZ* axes, respectively

Hence,

$$\bar{a} = \{\sin\varphi_{\rm a}\sin\alpha_{\rm K};\cos\varphi_{\rm a};\sin\varphi_{\rm a}\cos\alpha_{\rm K}\} \qquad (2.12)$$

The equation for line *A* crossing point **A** with directing vector **a** is of the form

$$\frac{x-r_t}{\sin\varphi_{\rm a}\sin\alpha_{\rm k}} = \frac{y}{\cos\varphi_{\rm a}} = \frac{z-z_t}{\sin\varphi_{\rm a}\cos\alpha_{\rm k}} \qquad (2.13)$$

The one-sheet hyperboloid is then described by the relation

$$\frac{x^2+y^2}{R^2} - \frac{z^2}{C^2} = 1 \qquad (2.14)$$

where *R* is the radius of the hyperboloid neck circumference taken as the shortest distance (along the common normal) between axis *Z* and any line on the hyperboloid surface.

Let us take the equation for the common normal to line *A* and axis *OZ*.

The parametric equation for the *OZ* axis is

$$x = 0 \qquad y = 0 \qquad z = t \qquad (2.15)$$

In vector form, the equation for the common normal is

$$(\bar{r}-\bar{r}_1)\bar{a}\cdot\bar{n} = 0 \qquad (\bar{r}-\bar{r}_2)\bar{z}\cdot\bar{n} = 0 \qquad (2.16)$$

where $\bar{r}$ = radius vector of the point on the common normal
$\bar{r}_1$ = radius vector of the point on line A
$\bar{r}_2$ = radius vector of the point on axis *OZ*
$\bar{a}$ = directing vector on line A
$\bar{z}$ = directing vector on axis *OZ*
$\bar{n}$ = directing vector of the common normal

Proceeding from the properties of the vector product, one can write $\bar{n} = \bar{a}\times\bar{z}$, where $\bar{z} = \{0;0;1\}$, and after substitution of $\bar{a}$ values from Eq. (2.12) and $\bar{z}$, we obtain

$$\bar{n} = \{\cos\varphi_{\rm a}; -\sin\varphi_{\rm a}\sin\alpha_k; 0\} \qquad (2.17)$$

Equations (2.16) in matrix form are

$$
\begin{vmatrix} x - r_t & y & z - z_t \\ \sin\varphi_{\mathrm{a}} \sin\alpha_{\mathrm{k}} & \cos\varphi_{\mathrm{a}} & \sin\varphi_{\mathrm{a}} \cos\alpha_{\mathrm{k}} \\ \cos\varphi_{\mathrm{a}} & -\sin\varphi_{\mathrm{a}}\ \sin\alpha_{\mathrm{k}} & 0 \end{vmatrix} = 0
$$
$$
\begin{vmatrix} x & y & z \\ 0 & 0 & 1 \\ \cos\varphi_{\mathrm{a}} & -\sin\varphi_{\mathrm{a}} \sin\alpha_{\mathrm{k}} & 0 \end{vmatrix} = 0 \tag{2.18}
$$

Upon transformations in Eq. (2.18), the matrices then take the form

$$
\begin{aligned}
&\sin^2\varphi_{\mathrm{a}} \sin\alpha_{\mathrm{k}} \cos\alpha_{\mathrm{k}} x + \cos\varphi_{\mathrm{a}} \sin\varphi_{\mathrm{a}} \cos\alpha_{\mathrm{k}} y \\
&\quad - (\cos^2\varphi_{\mathrm{a}} + \sin^2\varphi_{\mathrm{a}} \sin^2\alpha_{\mathrm{k}}) z \\
&\quad + \cos^2\varphi_{\mathrm{a}} z_t - \sin^2\varphi_{\mathrm{a}} \sin\alpha_{\mathrm{k}} \cos\alpha_{\mathrm{k}} r_t \\
&\quad + \sin^2\varphi_{\mathrm{a}} \sin^2\alpha_{\mathrm{k}} z_t = 0 \\
&\sin\varphi_{\mathrm{a}} \sin\alpha_{\mathrm{k}} x + \cos\varphi_{\mathrm{a}} y = 0
\end{aligned} \tag{2.19}
$$

Equations (2.19) define the common normal as an intersection of two planes.

Let us find an equation for the common normal in a symmetrical form. For this purpose, let us define an arbitrary point on the line and its directing cosines.

Let $x = 0$; it then follows from Eq. (2.19) that $y = 0$. If we substitute $x = 0$ and $y = 0$ in Eq. (2.19), we obtain, after transformations,

$$
z = \frac{\cos^2\varphi_{\mathrm{a}} z_t - \sin^2\varphi_{\mathrm{a}} \sin\alpha_{\mathrm{k}} \cos\alpha_{\mathrm{k}} r_t}{\cos^2\varphi_{\mathrm{a}} + \sin^2\varphi_{\mathrm{a}} \sin^2\alpha_{\mathrm{k}}} + \frac{\sin^2\varphi_{\mathrm{a}} \sin^2\alpha_{\mathrm{k}} z_t}{\cos^2\varphi_{\mathrm{a}} + \sin^2\varphi_{\mathrm{a}} \sin^2\alpha_{\mathrm{k}}} \tag{2.20}
$$

Thus, a point on the normal is singled out with the following coordinates:

$$
\begin{aligned}
x &= 0 \\
y &= 0 \\
z &= \frac{\cos^2\varphi_{\mathrm{a}} z_t - \sin^2\varphi_{\mathrm{a}} \sin\alpha_{\mathrm{k}} \cos\alpha_{\mathrm{k}} r_t}{\cos^2\varphi_{\mathrm{a}} + \sin^2\varphi_{\mathrm{a}} \sin^2\alpha_{\mathrm{k}}} + \frac{\sin^2\varphi_{\mathrm{a}} \sin^2\alpha_{\mathrm{k}} z_t}{\cos^2\varphi_{\mathrm{a}} + \sin^2\varphi_{\mathrm{a}} \sin^2\alpha_{\mathrm{k}}}
\end{aligned} \tag{2.21}
$$

The directing cosines of the common normal determined above are components of the vector **n** [see Eq. (2.17)]. Hence, the equations for the common normal in symmetrical form become

$$\frac{x}{\cos\varphi_{\mathrm{a}}} = \frac{y}{-\sin\varphi_{\mathrm{a}}\sin\alpha_{\mathrm{k}}}$$

$$z = \frac{\cos^2\varphi_{\mathrm{a}} z_t - \sin^2\varphi_{\mathrm{a}}\sin\alpha_{\mathrm{k}}\cos\alpha_{\mathrm{k}} r_t}{\cos^2\varphi_{\mathrm{a}} + \sin^2\varphi_{\mathrm{a}}\sin^2\alpha_{\mathrm{k}}} + \frac{\sin^2\varphi_{\mathrm{a}}\sin^2\alpha_{\mathrm{k}} z_t}{\cos^2\varphi_{\mathrm{a}} + \sin^2\varphi_{\mathrm{a}}\sin^2\alpha_{\mathrm{k}}} = z_0 \tag{2.22}$$

The point of intersection of the common normal and the Z axis is found from the following system of equations:

$$\begin{gathered} x = 0 \qquad y = 0 \qquad z = t \\ \frac{x}{\cos\varphi_{\mathrm{a}}} = \frac{y}{-\sin\varphi_{\mathrm{a}}\sin\alpha_{\mathrm{k}}} \qquad z = z_0 \end{gathered} \tag{2.23}$$

The point of line A and the common normal intersection is determined by the following equations:

$$\begin{gathered} \frac{x - r_t}{\sin\varphi_{\mathrm{a}}\sin\alpha_{\mathrm{k}}} = \frac{y}{\cos\varphi_{\mathrm{a}}} = \frac{z - z_t}{\sin\varphi_{\mathrm{a}}\cos\alpha_{\mathrm{k}}} \\ \frac{y}{\cos\varphi_{\mathrm{a}}} = \frac{z - z_t}{\sin\varphi_{\mathrm{a}}\cos\alpha_{\mathrm{k}}} \qquad z = z_0 \end{gathered} \tag{2.24}$$

As a result, we have

$$\begin{aligned} x &= r_t \cdot \left(1 - \frac{\sin^2\varphi_{\mathrm{a}}\sin^2\alpha_{\mathrm{k}}}{\cos^2\varphi_{\mathrm{a}} - \sin^2\varphi_{\mathrm{a}}\sin^2\alpha_{\mathrm{k}}}\right) \\ y &= -r_t \cdot \frac{\sin\varphi_{\mathrm{a}}\sin\alpha_{\mathrm{k}}\cos\varphi_{\mathrm{a}}}{\cos^2\varphi_{\mathrm{a}} + \sin^2\varphi_{\mathrm{a}}\sin^2\alpha_{\mathrm{k}}} \\ z &= z_0 \end{aligned} \tag{2.25}$$

The shortest distance between axis Z and line A (throat circumference radius of the hyperboloid) is the distance between the points calculated by Eqs. (2.24) and (2.25). Thus, coefficient R in the equation of hyperboloid (2.14) is defined as

$$R = r_t \frac{\cos\varphi_{\mathrm{a}}}{\cos^2\varphi_{\mathrm{a}} + \sin^2\varphi_{\mathrm{a}}\sin^2\alpha_{\mathrm{k}}} \times \sqrt{\cos^2\varphi_{\mathrm{a}} + \sin^2\varphi_{\mathrm{a}}\sin^2\alpha_{\mathrm{k}}} \tag{2.26}$$

The relation for the plane in which the throat circumference lies is $z = z_0$.

The equation for the free surface of braids is

$$\frac{x^2+y^2}{R^2}-\frac{(z-z_0)^2}{C^2}=1 \tag{2.27}$$

where R and z_0 are found in Eqs. (2.26) and (2.22), respectively.

For determining coefficient C in Eq. (2.27), the following argument is useful. Because line A lies on the hyperboloid surface, coordinates of any point considered as solutions of Eq. (2.13) will also solve Eq. (2.27). Hence, by substituting coordinates of the point on line A in Eq. (2.27), we obtain the equation for the relative coefficient C. However, to solve the problem of braiding a cylinder, and hence, finding coefficient C, with this aim, we will arrive at an identity. To avoid this, let us find a point on line A that is different from point **A** by transformation of Eq. (2.13) into parametric form

$$\frac{x-r_t}{\sin\varphi_{\mathrm{a}}\sin\alpha_{\mathrm{k}}}=t \qquad \frac{y}{\cos\varphi_{\mathrm{a}}}=t \qquad \frac{z-z_t}{\sin\varphi_{\mathrm{a}}\cos\alpha_{\mathrm{k}}}=t \tag{2.28}$$

We then define the coordinates of some point **A**. Let coordinate z of this point be z_t+1. This point then has the following coordinates:

$$x=\tan\alpha_{\mathrm{k}}+r_t \qquad y=\frac{1}{\tan\varphi_{\mathrm{a}}\cos\alpha_{\mathrm{k}}} \qquad z=z_t+1 \tag{2.29}$$

Coefficient C is found upon substitution of the marked point into Eq. (2.27)

$$C=\frac{z_B-z_O}{\sqrt{\frac{x_B^2+y_B^2}{R^2}}} \tag{2.30}$$

where x_B, y_B, and z_B are coordinates of the marked out point on line A.

The free surface of the braids is restricted by some (limiting) section of the mandrel from which the braids are coming off. It is also restricted by the guides of the braider along which the fiber carriers are moving. The location of circular guides relative the braiding center is characterized by some radius R_{m}. There exists a mutually identical correlation between the coordinate along axis Z of the limiting section and the location of the article relative to the braider when a layer of certain surface density is braided. If the coordinate of the limiting section is known, the respective location of the article dependent on coordinate Z_{m} can be determined based on the following system of equations:

$$\begin{aligned} &x^2+y^2=R_{\mathrm{m}}^2 \\ &\frac{x^2+y^2}{R^2}-\frac{(z-z_0)^2}{C^2}=1 \end{aligned} \tag{2.31}$$

Once the relationship between the points on the line of the braid lying over the mandrel surface and the article location is found, one can determine the required shifting rate of the article relative the braider in order to pack the braids with maximum surface density in the layer. In addition, it then becomes possible to estimate the structural parameters of the devices needed for installation of an intermediate spinneret and a unit for cross-unwinding.

2.3 Degree of Filling of Braided Layers

The geometric parameters of the braiding process (number of braids in each family, braiding angle, mandrel diameter, etc.) do affect the cross-sectional shape of braids in a layer. Their effect is exhibited in the constriction or widening of the braids during the braiding of a compacted layer. The cross-sectional width of a nonconstrained braid made up of several braids is defined by the empirical formula

$$T_S = 2.5 + \log_2 N_S \text{ [mm]} \tag{2.32}$$

where N_S is the number of folds of the initial braid and T_S is width of the resultant braid.

The geometric scheme of a braided layer can be constructed by taking interlacing braids that are rectangular in section along broken trajectories. The width of the braid cross-section can be calculated with Eq. (2.32) while the section height is found as follows:

$$\delta = \frac{N_S N_T \cdot 10^{-3}}{T_S \rho \cdot 0.907} \tag{2.33}$$

where N_T = initial braid size in tex (weight in gram of 1,000 meters of a fiber)

ρ = reinforcing fiber density

0.907 = filling degree of the initial braid with hexagonal packing (Fig. 2.4)

The rated dependencies based on the scheme proposed differed greatly from the results of experimental attempts to define the filling degree. Moreover, by proceeding from the above representation of the braiding layer, it is impossible to determine the boundary values of the geometric parameters that lead to increasing thickness due to overdense packing and other changes in the nominal sizes of structural elements.

In order to obtain the sought dependencies, one should bear in mind the real sectional shape of the reinforcing elements. The axial braids in the braided layer acquire a strongly extended ellipsoidal cross-section.

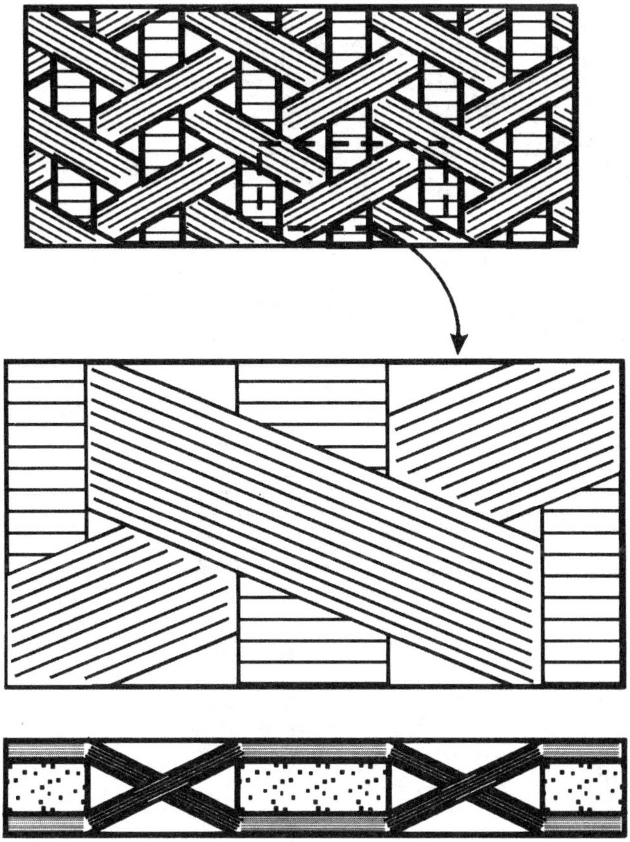

Figure 2.4 Interweaving scheme, disregarding the ellipticity of axial braids.

As a result, the formed layer has variable thickness and therefore can be more densely packed. If the ellipsoidal section is substituted with a polygonal section, an alternative scheme (Fig. 2.5) of interweaving can be constructed based on the initial scheme (Fig. 2.4). The axial braids in this scheme have a polygonal section, whereas the sleeving braids have a rectangular section. In this case, the polygonal section width and area are kept equal to the width and area of the rectangular section. It is believed that, for the substituting section, the following relation is true: $AB = t_0/4$.

Through the consideration of an elementary cell (Fig. 2.5), let us determine the maximum layer thickness

$$h = \delta_0 + 2x + \frac{2\delta}{\cos\alpha} \tag{2.34}$$

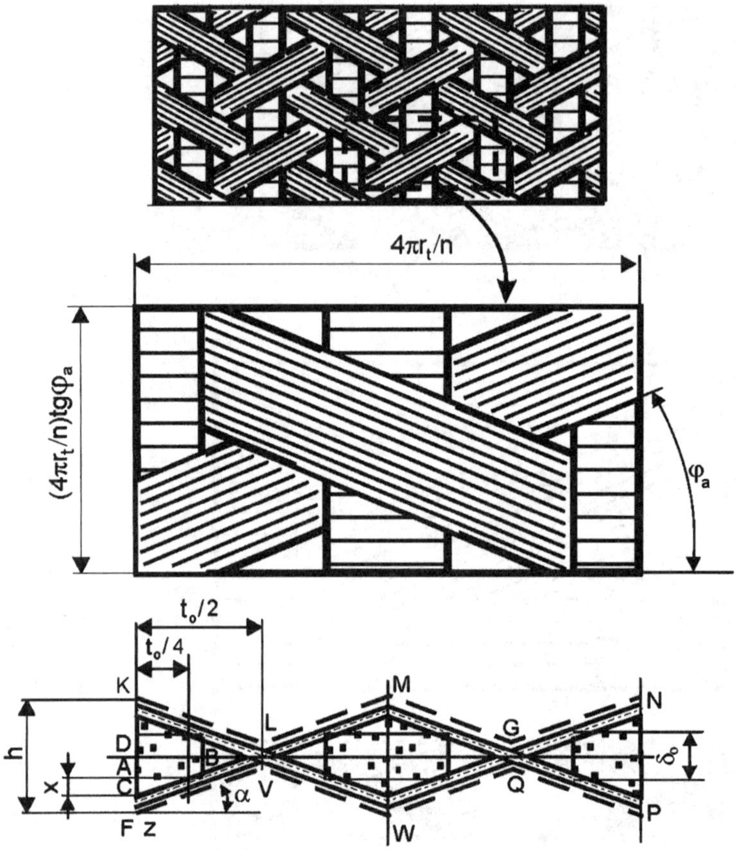

Figure 2.5 Filling degree dependence for the interweaving scheme, taking into account the ellipticity of the axial braids.

where δ_0 = sectional height of the axial braids
δ = sectional height of the sleeving braids
$\alpha = \arctan(4x/t_0)$

The cell face area (polygon KLMGNPQWVZ) is found from the relation

$$F_\mathrm{T} = h\frac{4\pi \cdot r_t}{n} - \left(\frac{4\pi \cdot r_t}{n}\right)^2 \frac{x}{t_0} \tag{2.35}$$

The elementary cell volume is

$$V = 2F_\mathrm{T}t_\mathrm{b} \cdot \cos\varphi_\mathrm{a} \tag{2.36}$$

The axial reinforcement volume is

$$V_{\mathrm{OC}} = \frac{8 t_0 \delta_0 \pi r_t \cdot \tan \varphi_{\mathrm{a}}}{n} \tag{2.37}$$

The braiding reinforcement volume is

$$V_{\mathrm{OP}} = \frac{t_{\mathrm{b}}}{\cos \varphi_{\mathrm{a}} \cos \alpha} \left(\frac{8 \pi r_t}{n} - \frac{t_{\mathrm{b}}}{2 \sin \varphi_{\mathrm{a}}} \right) \tag{2.38}$$

The distance between adjoining layers can be roughly estimated as

$$\Delta_{\mathrm{C}} = \frac{\delta_0}{2} 2 + x + \frac{\delta}{\cos \alpha}$$

The layer filling degree with the fibers will then be

$$\varphi_{\mathrm{f}} = \frac{V_{\mathrm{OC}} + V_{\mathrm{OP}}}{V} = \frac{\frac{8\pi r_t}{n \cos \varphi_{\mathrm{a}}} \left(t_0 \delta_0 + \frac{t_0 \delta_0}{\cos \alpha} \right) \frac{t_{\mathrm{b}}^2 \delta}{\sin 2\varphi_{\mathrm{a}} \cos \alpha}}{\left(\frac{4\pi r_t}{n} \right)^2 \tan \varphi_{\mathrm{a}} \left[\delta_0 + x \left(2 - \frac{4\pi r_t}{n t_0} \right) + \frac{2\delta}{\cos \alpha} \right]} \tag{2.39}$$

The value of x from the last formula, found from the similarity of triangles ACB and DEF, is

$$\frac{x}{t_0} = \frac{\frac{\delta_0}{2} + x + \frac{2\delta}{t_0} + \sqrt{x^2 + (t_0/4)^2}}{r_t \pi} \tag{2.40}$$

From this relation, the dependence on x as a positive root is obtained

$$x^2 \left[\left(\frac{\pi r_t}{n} - \frac{t_0}{4} \right)^2 - \frac{\delta^2}{4} \right] - \frac{t_0 \delta_0}{4} \left(\frac{\pi r_t}{n} - \frac{t_0}{4} \right) x + \frac{t_0^2}{64} \left(\delta_0^2 - \delta^2 \right) = 0 \tag{2.41}$$

The interrelation of structural parameters at which $x \geq \delta_0/2$ can be interpreted as an overcompacted state of the braided layer. The results of calculations by (2.41) are consistent with experimental data within a wide range of various structural parameters of the layer. The calculation results obtained agree well with experimental values (Table 2.1).

TABLE 2.1 Theoretical (Numerator) and Experimental (Denominator) Results of Determining the Filling Degree of the Braided Layer

	φ_a, degrees				
t_b/t_0	15	25	35	40	50
4/1	—	0.659/0.62	0.614/0.6	0.605/0.6	0.596/0.58
2/1	0.75/0.7	0.606/0.6	0.575/0.55	0.568/0.53	0.562/0.5
1/1	0.663/0.62	0.564/0.54	0.545/0.52	0.533/0.5	0.525/0.5
1/2	—	0.574/0.55	0.534/0.5	0.572/0.5	0.512/0.48
1/4	—	0.621/0.58	0.538/0.5	0.501/0.5	0.486/0.48

For a layer reinforced by a 2D scheme, the filling degree can be calculated by the relation

$$\varphi_f = \frac{\varphi_1 \cdot t_b}{\frac{4\pi r_t}{n} \sin \varphi_a + t_b} \tag{2.42}$$

where φ_1 is the filling degree of the initial braid.

Note that the limiting values of the filling degree can be obtained with account of the results cited in Sec. 2.1. In this case, each reinforcement angle will correspond to a certain strip width of the reinforcing material. Proceeding from these data, the filling degrees of braided layers are presented in Fig. 2.6, where each reinforcement angle corresponds to a certain strip width. Apparently, the material should be overcompacted to increase the filling degree, which, however, may widen the outer diameter of the article. The reinforcing angle will inevitably change in the spinneret during molding of such an article, and a considerable increment in its diameter will result in a wave in the material that

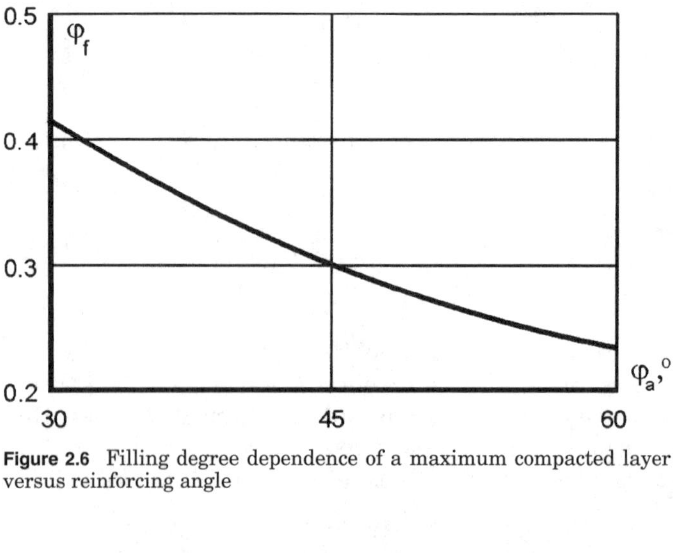

Figure 2.6 Filling degree dependence of a maximum compacted layer versus reinforcing angle

will augment the pulling force and eventually clog the article in the spinneret.

2.4 Conclusions

1. From the analysis of the data presented, the conditions for forming a braided layer of a given structure have been determined. In addition, the peculiarities of automatic braiding have been established, based on which a calculation method for technological regimes and a methodology for selecting rational, constructive decisions for the members of necessary equipment have been derived.
2. From the analysis of the free surface of braids, dependencies have been obtained for determining the boundary-limiting planes, the required pulling velocity of items for the respective numbers of carriers, and the rotation velocities of spindles.
3. From the analysis of the filling degree of braided layers, it has been made possible for us to determine the threshold values of the filling degree to avoid overcompaction of the structure and to elaborate the optimum technological regimes of pulling.

Chapter

3

Pultrusion Regimes

3.1 Requirements of Pultrusion Components of Polymer Composites

The choice of reinforcing fillers and binders and their relationship, reinforcing schemes, and interactions at the interfaces between components depend upon the technological process used and the purpose of the final items. The components acquire certain characteristics from their individual elements, which comprise the needed technological properties of the material as a whole, just as the separate production stages also contribute to the properties of the final items.[7,31,32,34,37,260]

To make pultrusion stable and reproducible, it is important to find a binder whose technical characteristics,[132,208,318] including variable rheological properties at different stages and compatibility with the reinforcing filler, are optimal for a given molding regime. In addition, the binder mainly governs the performance of the pultrusion materials, namely their continuity, resistance to cracking and atmospheric effects, bearing capacity, and so on.

Thermosetting binders are most commonly employed in pultrusion of profile articles, along with thermoplastic matrices that can increase pultrusion capability in manufacturing reinforced plastics.[94,384] Use of thermosetting binders makes it possible to vary the technological properties of goods produced,[58,207,208] namely their viscosity, wettability, and degree of impregnation during combination with the reinforcing filler, preshaping, and final molding of profile articles. This variability can be attained through regulation of the initial structure, molecular mass, and ratio of the main components, as well as by introduction of various modifiers. The problems that arise most often during pultrusion of fibrous thermosetting polymer composites include solidification reactions of the binder during material and article molding, changes in rheological properties during setting, volume shrinkage, and exothermal and other effects.

Unsaturated polyester resins that could ensure low viscosity and durability during initial stages and a high setting rate devoid of isolation of low-molecular matter under relatively low temperatures within the range 90 to 150°C turned out to be most appropriate in composite pultrusion. Among drawbacks of these binders, however, are their rather low strength and the toughness of final products whose heat-induced strain temperature is below 100°C, which explains why they are scarcely used in production of structural materials.

Phenol formaldehyde or polyamide thermosetting binders may impart high heat resistance to the composites; however, the former liberate large amounts of volatile matter at setting, and the latter require high curing temperatures.

The optimum combination of strength, toughness, and heat-resistance properties of thermosetting composites requires the use of epoxide binders, such as difunctional resins—commonly dian—or polyfunctional resins that set without liberating volatile products and shrinking slightly at relatively low temperatures. At setting, they form heat-resistant, densely netted polymers with high physical and mechanical properties. Nevertheless, epoxide binders are most often used, and they often contain amine hardeners with high original viscosity that necessitate either active diluters or elevated temperatures to effectively impregnate the fibrous fillers. Such binders can hardly fulfill the requirements of prolonged durability and high setting rate during pultrusion. Preference is most often given to epoxy binders based on liquid anhydride hardeners in combination with curing catalysts. The catalysts display low original viscosity, relatively high tenacity in normal conditions, and high setting rate at elevated temperatures. To improve their processing and operation characteristics, such binders can be modified by different additives.

The properties and structures of epoxide binders in their original state, during curing under various conditions, and their degree of transformation in the solidified state, as well as different chemical and physical setting methods, have been described at length in the literature, as have their setting kinetics.[98,127,143,149] It has been proven that the chemical and related physical (phase and relaxation) transformations of the binders during setting are strongly interrelated, with the mechanism and kinetics of these transformations, structure, and properties of the forming polymers being highly sensitive to changes in setting conditions and binder composition. The capabilities of applying these binders in pultrusion processes can be verified by the chemorheological state diagrams (TTT diagrams),[127] which demonstrate the mechanism of gelation and glass transition of the binders.

The choice of fibrous filler, its volume content, and reinforcing scheme at pultrusion is evidently specified by urgency in reaching the set level of performance on the one hand and resistance to breakage at pulling on the other. With this aim, mainly continuous fillers such as roves, braids, bands, or fabrics, are employed at volume contents from 40 to 80 percent. To a larger extent, glass fiber fillers are used because carbon fiber-based fillers present certain difficulties due to their toughness and brittleness. In a number of cases,[193,198,204,230,921] hybrid fillers based on different fibers are used to attain high-strength and high-quality products.

During production of composite profile articles, the reinforcing filler is laid along the article axis to add mechanical strength in this direction. For transverse reinforcing, fabrics, mats, woven strips, or continuous multidirectional reinforcing fillers based on glass, carbon, and aramide fibers are often used. Similarly to the cases presented above, fillers based on 3D textures ensure isotropy of mechanical characteristics of the profile articles and improve their shear strength while preserving longitudinal strength.

Apart from the toughness and strength characteristics of the materials in the pulling direction, the filler origin, specific area, treatment type, and fiber packing density in braids as well as that of the braids are also of importance in the material. These characteristics are critical in defining the wettability and degree of saturation by the initial binder as well as filler cohesion with the polymer matrix and degree of reinforcement. The strength and toughness of the final article are also conditioned by adhesive interactions at the interfaces between components. It is stated in a number of works[35,85,117,299,380,404,409,416,801] that the strength of composite materials at transverse loading, compression, and shear is dependent upon the interfacial strength. To derive a solid material, the relation of adhesive strength τ_a to that of the filler σ_f should be within limits of $\tau_a/\sigma_f \geq 0.015$ to 0.040.

Analysis of various components for pultrusion of composite materials has shown that, from the viewpoint of processability and imparting specific properties in continuous production processes, the most optimal are those composites that use an epoxide binder base with different fibrous fillers.

3.2 Parameters of the Pultrusion Process

Pultrusion includes a number of main stages, namely filler impregnation with the initial binder, molding of a semifinished workpiece, molding of an article, fixing of its shape, and final curing of the binder. Except for the latter, which can be run separately, all of these stages are united to form an uninterrupted, rapid process. Due to this fact, it is difficult to select and optimize pultrusion regimes that proceed from the known component characteristics and quality demands to the final articles while also optimizing process productivity, energy consumption, and so on.

Pultrusion has been comprehensively studied in several works.[23,76,164] For example, Ref. 70 presents a model of a technological process: data on optimizing the process of manufacturing items with high dynamic characteristics are cited, and the effect of pultrusion parameters on the articles produced has been considered.[119,206,209] Studies have also been published that investigate the individual parameters and stages of the process, including impregnation,[118,126,296,298] methods of binder introduction,[207,358] effects of the spinneret lead-in,[119,317,319] influence of pressure in the binder at the spinneret inlet,[131,297,318] binder flows in the molding process,[131,133] effect of pulling velocity,[68,71] types of heater,[214] temperature conditions,[49,68,132,303,372,373,415] and calculations of pulling force.[205]

The impregnation stage of the reinforcing filler by thermosetting binders in a saturation bath differs little from that in the similar process of wet winding or manufacture of durable prepregs, including those based on epoxy binders.[208]

The dominating effect on the degree and uniformity of filler saturation as well as the total covering of the binder is exerted by the viscosity and surface tension of the polymer composition, filler porosity, and its surface treatment, pulling velocity, and tension value. The stage of pultrusion during which a needed binder covering is provided is determined. Impregnation terminates as soon as the required filler-to-binder ratio is reached, and excess binder is removed during the preforming and final molding of the article.

The principal stage of pultrusion that governs its rate and quality of items produced is molding of the profile article in a heated molding spinneret. Here, complex physicochemical interactions take place while the profile is molded and its shape is fixed as the binder sets to a specified degree. The process rate is dependent on reactivity of the polymer composition, being about 0.2 to 0.5 m/min for the epoxide–anhydride binders.[67,385] Binders, in particular polyester–epoxide binders, permit a pulling velocity of 0.7 to 0.9 m/min. The variable technological parameters of this stage are the temperature field distribution in the forming spinneret and the pulling velocity (setting time). Together with the spinneret length, these parameters assign the time–temperature conditions of molding and article shape fixing and affect the pulling force. They are interrelated; they are connected with the processing properties of the material and their changes during binder curing. Detecting the relationship of these factors is an extremely complex problem, and it is critical in optimizing the molding regimes (spinneret temperature, velocity, and force of pulling, etc.). Solution of the problem requires that heat exchange between the pulled profile, spinneret, and ambient space be taken into account in the case of exothermal binder curing behavior.[415] In addition, it is necessary to take into account the plastic deformation of the material at pulling; the changes in its properties and volume in response to thermal effects and binder curing; and the variations in shape, length, and area; and the quality of the forming channel. Of course, it is impossible to consider all the factors, therefore certain simplified analytical (semiempirical) models[68,131,148,164,228,318,334,372,414] or purely empirical investigation methods of individual factors[67,132,176,317,336,337,385,473] are used.

To abbreviate the duration of the process and cut energy consumption during molding in the heated spinneret, the binder is not fully set but ensures fixing of the article form during pulling and cutting. To reach the full setting degree of the binder that promotes the required deformation, strength, and thermal properties, the cut profiles are aftercured.

Rheological characteristics of the binder at curing depend much on curing temperature and thermal effects of the curing reaction. As a result of chemical transformations, the forming polymer molecular mass and heat-induced mobility both change due to growing and branching chains and due to the changed concentration of physical and chemical bonds.[24,63] The dominating effects here are the gelation reaction, during which a continuous polymer net is formed and a glass transition occurs that freezes the segmental thermal displacement and makes the binder viscosity infinitely high. An irreversible transition from a viscous flow to a viscoelastic state takes place during gelation, whereas at glass transition a reversible conversion from a viscous flow into a glassy state occurs. The processes considered above are observed in chemorheological state diagrams. TTT diagrams involve the kinetic data presented in temperature–time coordinates (Fig. 3.1) and equilibrium in temperature-setting degree coordinates (Fig. 3.2).[127,268]

Analysis of TTT diagrams forms a basis for a substantiated choice of curing regimes in the pultrusion spinneret. To eliminate profile deformation at the outlet, its curing should lead to glass transition of the binder after gelation. To attain a high degree of cure, final curing should be carried out in a separate heat aggregate under the temperatures that approach the glass transition of the set composition (T_{GE}).

As the study of available information proves, the main curing reaction proceeds within the temperature range 80 to 200°C. Introduction of various additives into the binder may essentially effect the setting kinetics and degree of the binder hardening. When an anti-adhesive additive such as zinc stearate is impregnated, the curing time changes significantly (Fig. 3.3). Introduction of this additive at 1 to 1.5 mass parts per 100 mass parts of epoxy resin improves pultrusion conditions

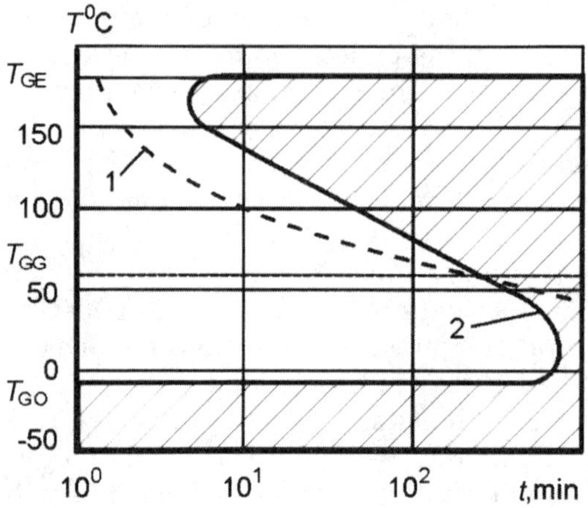

Figure 3.1 Diagram of chemorheological states and transformations of epoxy binder: (1) gelation; (2) glass transition. Glass-transition temperatures of the solidifying composition: unset (T_{GO}); forming gel (T_{GG}); set (T_{GE}).

and reduces pulling force.

In the viscous-flow state, during which the binder is combined with the filler, it is critical to know the absolute values of the binder's real viscosity. As it follows from Fig. 3.4, the initial viscosity of the binder at $t \leq 60°C$ and its variation in time comply with the processing requirements concerning degree of impregnation and tenacity.

At higher temperatures (Fig. 3.5), a competitive effect between the

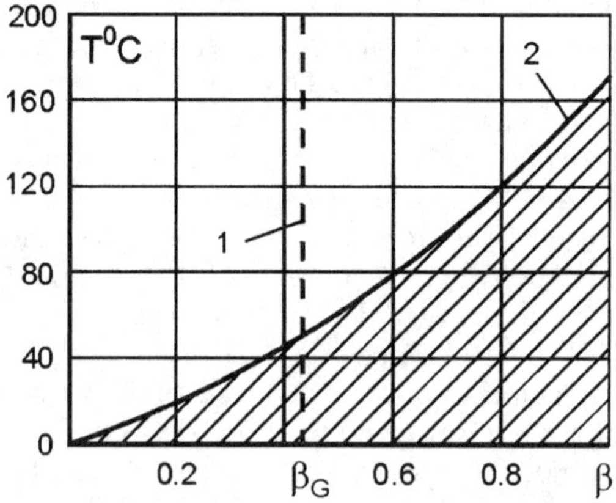

Figure 3.2 Equilibrium diagram of chemorheological states and transformations of epoxy binder: (1) gelation; (2) glass transition.

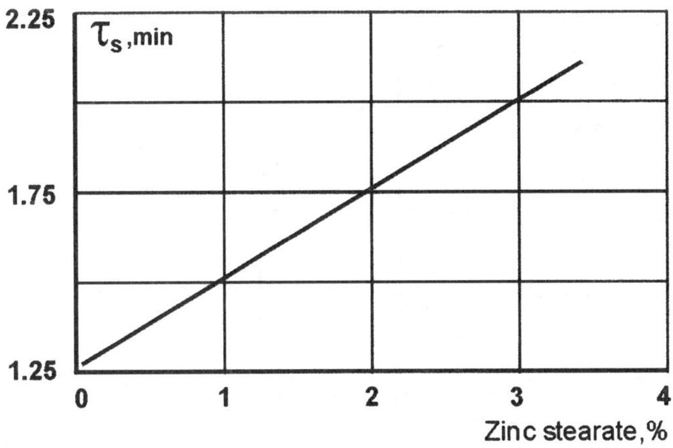

Figure 3.3 Curing time to binder hardening degree $\beta = 0.8$ versus zinc stearate content.

two phenomena is observed, namely a drop in viscosity due to heating and an increase in viscosity due to rapid curing. The former phenomenon is most clearly observed under the temperature range 100 to 120°C, while the latter gains strength within the range 140 to 180°C.

This phenomenon plays an important role in the choice of time–temperature regimes of pultrusion when the heating rate increases to 150 to 200°C/min. Thus during the pultrusion process for composite products, the range of pulling velocities varied from 0.2 to 0.8 m/min and the maximum temperature in the spinneret ranged from 150 to 250°C.

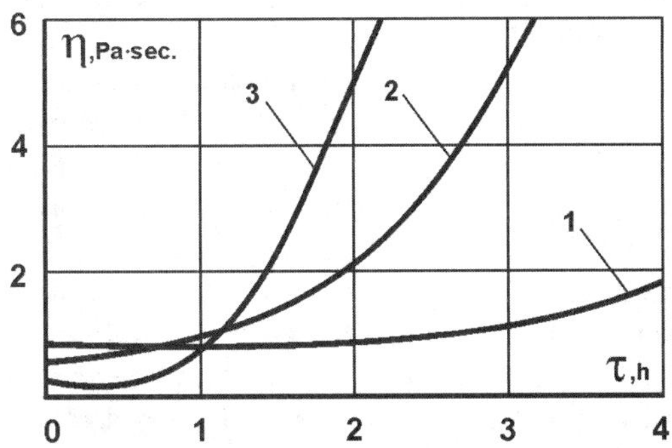

Figure 3.4 Variations in epoxy binder viscosity within the temperature range 20 to 55°C: (1) 20°C; (2) 40°C; (3) 55°C.

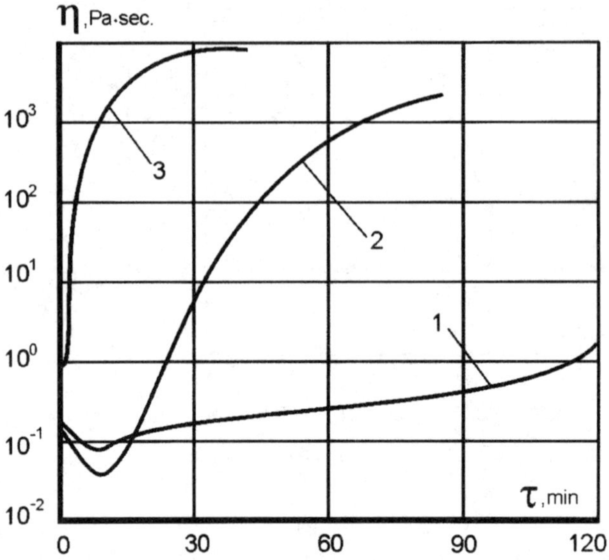

Figure 3.5 Variations of binder viscosity within the temperature range 60–160°C: (1) 60°C; (2) 100°C; (3) 160°C.

3.3 Pultrusion Process Analysis in the Heated Spinneret

The principal stage of pultrusion of thermosetting composites is pulling of the reinforcing filler impregnated with a curing agent through the heated spinneret. The fibrous filler is compacted in the spinneret, the binder excess is removed and redistributed within the material, and the product is cured to setting to the required degree of physicomechanical properties. As a result of the curing reaction and temperature alteration, the binder experiences both volume expansion and shrinkage. It also undergoes physical (phase and relaxation) transformations as well as a transition from the viscous-flow state via jelly-like, elastic, and then solid glassy states. The relationships of the thermal and kinetic parameters of pultrusion, including pulling velocity, binder curing rate, and temperature field distribution in the forming spinneret, dominate this stage of pultrusion. In turn, the temperature field distribution in the spinneret is dependent upon the working channel length and configuration, the contact area and characteristic dimensions of the article, its structural parameters, the thermophysical and adhesive properties of components, and the magnitude of the exothermal effect at curing.

To estimate the main parameters of the pultrusion process while taking into account the chemical and physical reactions that occur at the setting stage, a series of mathematical models has been elaborated based on the experimental evidence of the temperature distribution along the working channel and heat exchange between the article and the spinneret walls.[23,148,158,188,414] Computations by these models are rather intricate and involve large volumes of experimental data. For thin-walled articles, these models can be simplified essentially by presuming that the heat flow occurs only in one direction and that the other two directions can be neglected. In this way, a system of two equations is made, consisting of a kinetic equation for the binder curing[309] and a univariate atypical equation of heat conductivity and heat transfer for the profile–spinneret system.

$$\frac{\partial \beta}{\partial t} = F(\beta) \cdot e^{-E(\beta)/RT(t)} \tag{3.1}$$

$$\rho C_{\mathrm{p}} S \frac{\partial T}{\partial t} - \lambda S \frac{\partial^2 T}{\partial x^2} + \rho C_{\mathrm{p}} S V_S \frac{\partial T}{\partial x} + hP(T - T_E) - Q_E \rho S \frac{\partial \beta}{\partial t} = 0 \tag{3.2}$$

where ρ = profile material density
C_{P} = heat capacity of profile materia
S = cross-sectional area of profile
P = perimeter of profile
λ = K-factor of the profile
V_S = pulling velocity
h = heat transfer factor of the profile medium
T = profile temperature
T_{E} = heat-exchange medium temperature
t = time
x = univariate system coordinate
Q_{E} = complete calorific effect of curing reaction
β = degree of binder setting

For this model the following experimental data are needed: kinetic parameters of non-isothermal curing of the binder within the material; function $F(\beta)$ and activation energy $E(\beta)$; and the temperature field distribution along the spinneret. The kinetic parameters of non-isothermal curing are usually estimated by the differential scanning calorimetry.[196] Using the derived kinetic curves of curing at different curing rates, one can determine the function $F(\beta)$ and activation energy $E(\beta)$ by Eq. (3.1).

The model takes into account the exothermal reactions of curing and variations in the thermophysical properties of the profile material. In the simplest case, the solution of the heat conductivity problem is based on the average thermophysical characteristics of the setting depth, independent of temperature.

The general methodology of selecting pultrusion regimes can then be formulated as follows. The relationship between the curing time, temperature, and setting degree can be found for a given binder from chemorheological diagrams. The criterion of selection is the necessity for the binder to transfer beyond the gelation area, in which case the glass-transition temperature should be above that of the article to a distance of the portion cut on leaving the spinneret. Based on data from the temperature field distribution in the spinneret at a given temperature and the thermokinetic parameters of the binder curing, the interrelation of the spinneret heating temperature to pulling velocity is calculated to reach the minimal setting degree of the article at the spinneret outlet. In the calculations, either general dependencies[23,148,158,188,414] or a simplified univariate model, as described above for the case of thin-walled items (wall thickness under 10 mm), can be used.

As an example, we present the investigation results of a pultrusion process for a tubular rod with 19 mm outer diameter and 1.5 mm wall thickness. For the binder, a composition based on a dian epoxy resin and isomethyltetrahydrophthalic anhydride has been chosen in combination with tertiary amine as a hardener. The described binder displays low initial viscosities of 232 Pa s (20°C), 46 Pa s (40°C), and 11 Pa s (60°C). Its elasticity modulus and tensile strength at room temperature are 3.2 GPa and 76 MPa, respectively. The glass-transition temperature is 120°C.[373] Zinc stearate has been used as an antitack agent.

To examine temperature variations in the forming spinneret and directly in the article being pulled, a steel spinneret of length 500 mm and a mandrel with a cross-sectional diameter corresponding to that of the article have been employed. The profile temperature has been controlled by a thermocouple placed between longitudinal braids and passing through all heating zones of the spinneret. To measure temperature at separate points along the working channel, the thermocouple was installed within a bunch of dry longitudinal reinforcing fibers. Readings have been taken in certain points during pulling until the temperature reached a constant value. The temperature in the bunch of dry reinforcing fibers has been taken as equal to that of the working channel when the pulling velocity $V_S = 0$.

The spinneret was heated using four 300 × 80 mm flat heaters of 1 kW power each. By altering the location of the heaters along the spinneret length, the temperature field distribution can be changed specifically. Considering the structural features of the forming unit, the heaters are installed at three points, namely near the inlet and outlet of the spinneret and at the middle (Fig. 3.6).

As experiments have shown, the temperature field is distributed lengthwise along the spinneret channel following a parabolic law with a small plateau at a distance of 25 to 30 cm from the inlet, which

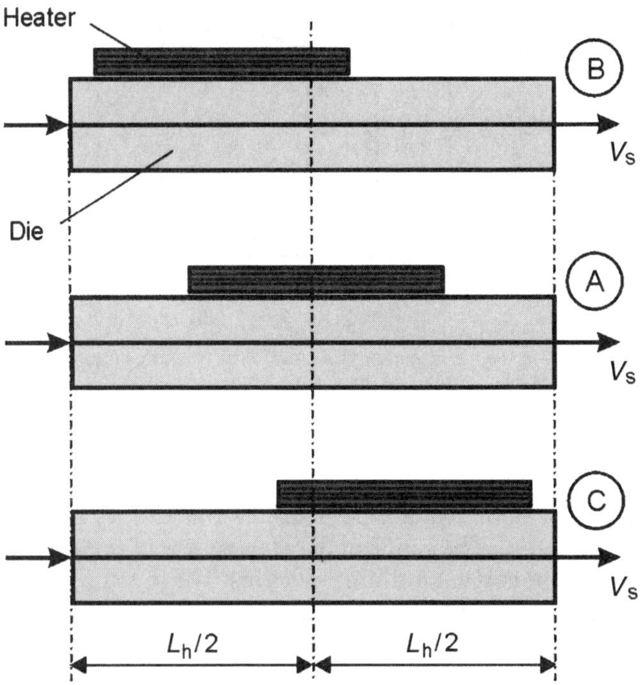

Figure 3.6 Layout of heaters lengthwise along the forming spinneret.

eliminates dependence on heater location (Fig. 3.7) and, independently, on velocity of the filler motion. The relationship of the pulling velocity of the dry reinforcement and the temperature field variation over the spinneret length is shown in Fig. 3.8.

It is evident that, with increasing pulling velocity, the maximum temperature peak shifts to the outlet, irrespective of heater location. In

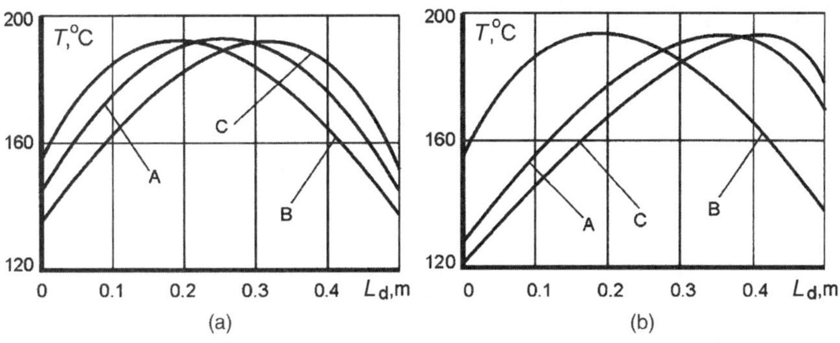

Figure 3.7 Temperature distribution along the forming spinneret at different heater arrangements, $T_H = 200°C$: (*a*) $V_S = 0$; (*b*) $V_S = 0.4$ m/min.

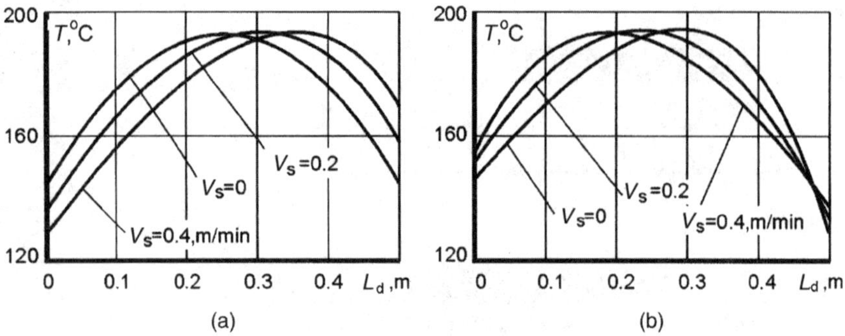

Figure 3.8 Temperature variation lengthwise along the spinneret working channel, depending on the pulling velocity of dry filler at location of heaters: (*a*) A; (*b*) B and $T_H = 200°C$.

contrast, the temperature difference at the spinneret inlet and outlet and, consequently, the dynamics of temperature increase, vary significantly. Hence, a conclusion can be made on the feasibility of regulating polymerization regimes for certain pulling velocities. This circumstance has been employed earlier in pulling articles through long spinnerets with set temperature regimes in given zones.[87]

To estimate the effect of the setting binder on temperature distribution in the working channel, i.e., in the actual composite material, it seemed expedient to study temperature variation using a thermocouple placed in the material upon impregnation.

The results obtained are presented in Figs. 3.9 and 3.10. The temperature distribution behavior is preserved, and the maximum temperature in the material is 7 to 8°C higher than that of the working channel

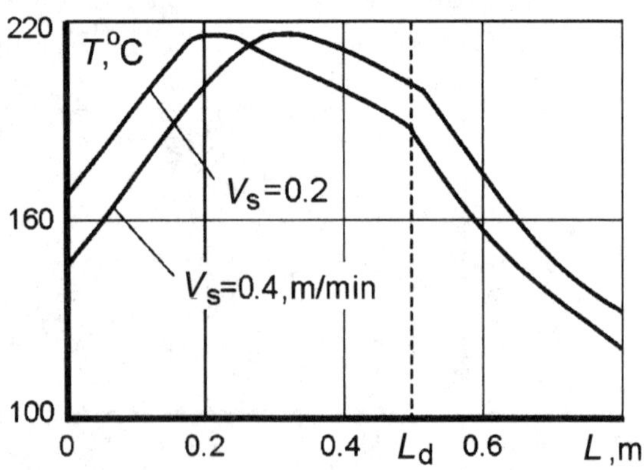

Figure 3.9 Temperature variations in the composite over the length of the working channel at different pulling velocities with heaters in position A and $T_H = 200°C$.

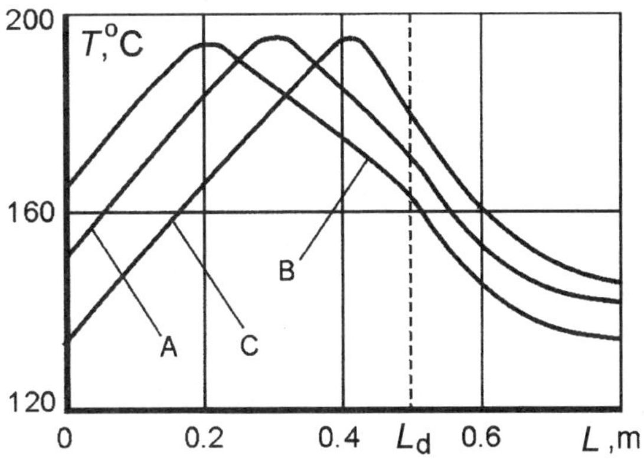

Figure 3.10 Effect of heater arrangement lengthwise along the spinneret upon temperature variation in the composite at $T_H = 200°C$ and $V_S = 0.4$ m/min.

walls. This can be attributed to the exothermal reaction of the binder polymerization.

The curing kinetics of the binder has been studied by differential scanning calorimetry. As a result, a series of kinetic curves of curing has been drawn at different heating rates. Then, based on Eq. (3.1), function $F(\beta)$ and energy $E(\beta)$ can be calculated. In Fig. 3.11, results of the calculations are presented for the degree of binder setting lengthwise along the spinneret as a function of temperature. The dotted lines in the figure correspond to the experimental values of the degree of polymerization of a profile ready in the outlet.

The degree of binder setting as a function of the pulling velocity and heater location(s) has been determined to optimize the choice of thermal

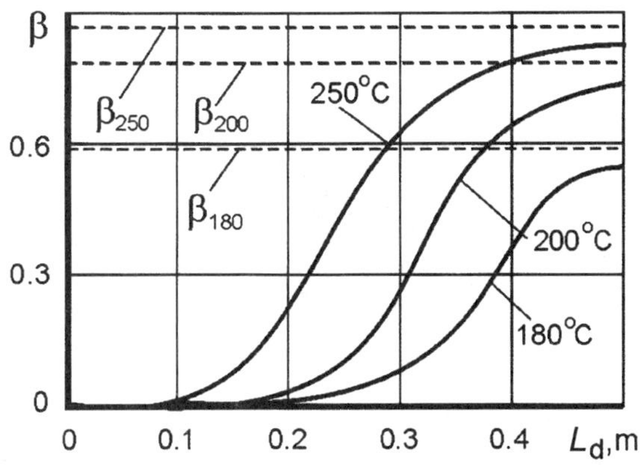

Figure 3.11 Rated values of the degree of binder setting versus the setting temperature.

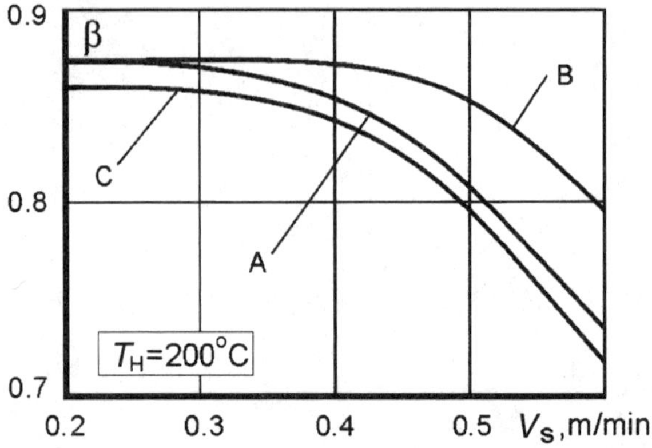

Figure 3.12 Changes in the degree of binder setting in glass-reinforced plastic tubular rods upon leaving the spinneret as a function of heating conditions and pulling velocity.

treatment regimes. The derived results are shown in Fig. 3.12. It is evident that, by placing the heaters closer to the spinneret inlet, the setting degree decreases somewhat at low pulling velocities. If the pulling velocity is high enough, the setting degree decreases drastically.

Because of the results obtained, experimental investigations were carried out in order to estimate the dependence of the degree of setting on pulling velocity and pultrusion temperature conditions. Experimental evidence is presented in Fig. 3.13.

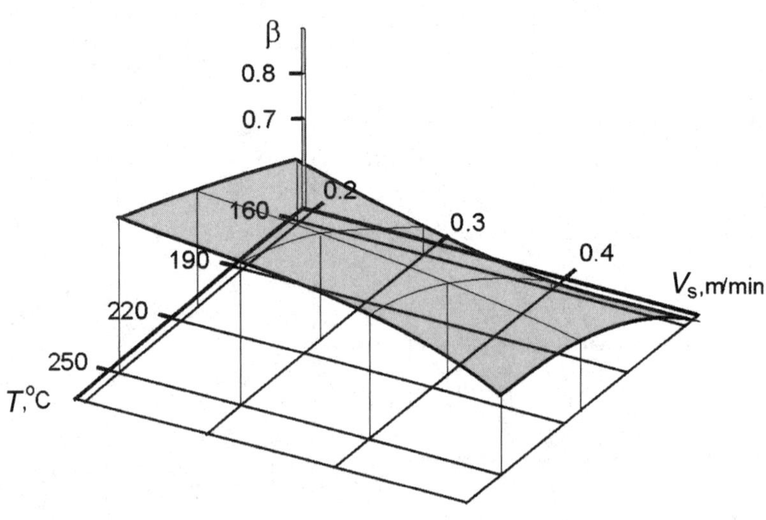

Figure 3.13 Variation in degree of setting of a ready profile versus pulling velocity and heater temperature.

It can be concluded, on the basis of the analysis presented, that the required setting degree of the binder can be obtained in the material by selecting the corresponding pulling velocity and temperature in the spinneret. Note that a temperature increase from 220 to 250°C does not noticeably increase the setting degree. However, an increase greater than 0.4 m/min in the pulling velocity leads to adverse results, while a decrease to 0.4–0.5 in the setting degree makes pulling intractable.

3.4 Conclusions

1. From analysis of components used for pultrusion of profile products, it has been established that, with respect to processing ability and obtaining certain properties under the conditions of a continuous technological process, it is expedient to use epoxy binders and different fibrous fillers.
2. From investigations of the choice of technological parameters for the pultrusion process, it is now possible to estimate the velocity and temperature limits, which are to be within 0.2 to 0.5 m/min and under 150 to 250°C in the spinneret.
3. From investigations conducted to study the effect of the temperature distribution in the spinneret on the setting degree of the material, optimum heater temperatures fall within the range 200 to 240°C and the optimum pulling velocity approaches 0.5 m/min when heaters are installed near the spinneret inlet.

Chapter

4

Investigation and Synthesis of Principal Structural Parameters of Complex-Profile Composite Ware

4.1 Structure and Structural Strains in Composite Profiles

The design and manufacturing technology of composite ware are interrelated with the choice of one of the two force-scheme concepts:[85,121,153,182]

1. The synthesis or the integral principle of design, during which reinforcing fibers are arranged to furnish the structure with the best resistance to all accepted types of effects. The integral composite structure is made of various elements by concurrent molding in one technological cycle.
2. The differential design principle presupposes formation of the force system of elements having distinct functional purposes, each of which is sensitive to one or several definite standard effects but is practically indifferent to other types of loads.

The differential principle of traditional technologies presumes multistage production of composite articles. As a result, the article accumulates a perceptible amount of defects and, consequently, great scatter in strength properties, leading to complications in design, computation, and expansion of load factor span. Both integral and differential methods are realized by optimum nonequivalent by mass structures.

Because loaded profile elements are, as a rule, thin-walled structures, it is difficult to keep to the synthesizing principle. It is difficult to alter reinforcement schemes (as required by the synthesizing principle) in thin, bearing layers as alterations involve increased labor, an increased occurrence of defects in junctions, and an increase in the mass of the structure. So far, the synthesizing principle is commonly employed for averaged force fields of the loads within the bounds of the structure. The reinforcement scheme is taken in this case to be invariable for the whole structure and is chosen in terms of the most loaded zone.

A great number of rod products with force purposes experience the effects of complex bending and combined loads, bringing about considerable normal and tangential stresses perceived by different elements of the profile article (Fig. 1.3). The profile configuration in this case should assist in reaching the optimum correlation between its bearing capacity and mass characteristics.

Thin-walled structures of profiles that have complex cross sections are characterized by available areas of just as positive or negative curvature. The most applicable profile-types rod elements produced with composite materials (CM) are the double-T shape.

A typical design of the profile force element commonly includes three main types of structural elements: flanges, walls, and connecting elements, such as shoulders. As all structural elements of profiles are formed, in principle, in one working cycle, their connecting units are included in the layers of both flanges and walls (Fig. 4.1).

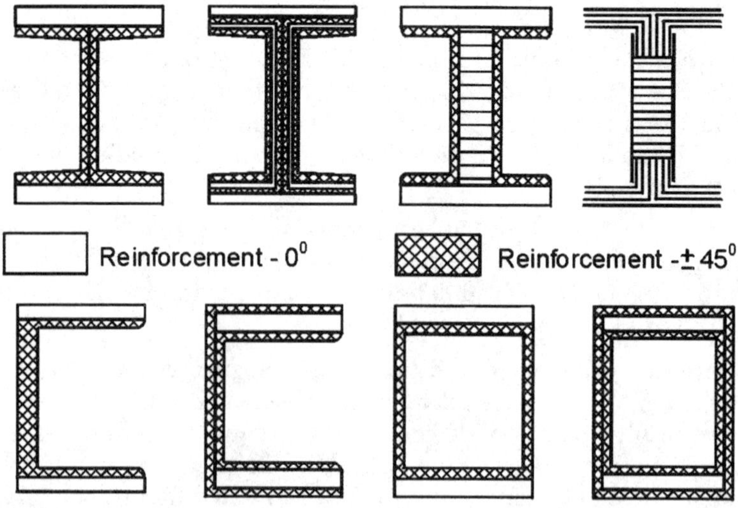

Figure 4.1 Schemes of beams of laminated CM.

During operation, the different elements of complex-profile composite articles experience different external loads. The flanges largely perceive longitudinal compression and tension stresses of bending moments, while the walls experience shear stresses of intersecting forces. The articles differ accordingly in structure and physic-mechanical characteristics. Thus far, the flanges are predominantly reinforced at 0° and the walls at ±45° to the longitudinal axis of the article.

This explains why certain additional stresses arise at the interface and inside the elements because of differences between the linear thermal expansion factors (LTEF) and Poisson ratios at joint deformation. Therefore, one of the fundamental principles of designing composite ware is application of only heat-balanced reinforcement schemes.[45,136]

As stated,[42,44,45,79,89,90,100,285,418] some laminated composite structures, including those studied herein, show a tendency for breakage through delaminating that starts on outer unloaded edges as a result of interlayer tangential stresses. In profiles, this delaminating occurs at the flange–shoulder boundary.

Edge effects in composite beams have been analyzed[180,262] using the model presented in Fig. 4.2. According to this model, significant tangential stresses appear near the free flange end, whose value at cooling from polymerization temperature T_{p}^{0} down to the operation temperature T_{m}^{0} can be found by the formula

$$\tau_{zyT} = \left(T_{\mathrm{m}}^{0} - T_{\mathrm{p}}^{0}\right)(\alpha_{zs} - \alpha_{zf})\frac{1}{k\Psi}\frac{\cosh kx - \cosh k(b - x)}{\sinh kb} \tag{4.1}$$

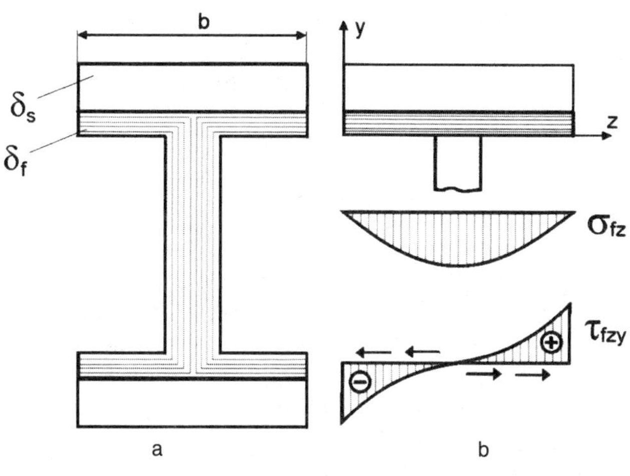

Figure 4.2 Edge effects in complex-profile composite articles: (*a*) profile structure; (*b*) computation model and stress distribution in shoulders.

where α_{zs} and α_{zf} are the LTEF of the flange and shoulders along the z axis,

$$k^2 = \frac{E_{zs}\delta_s + E_{zf}\delta_f}{\Psi E_{zs}\delta_s E_{zf}\delta_f}$$

$$\Psi = \frac{\delta_s}{2G_{zys}} + \frac{\delta_f}{2G_{zyf}} \tag{4.2}$$

E_{zs}, E_{zf}, G_{zys}, G_{ztf}, δ_s, and δ_f are elasticity, shear, and thickness module of the shoulders and flange, depending on CM structure.

Normal stresses in shoulders are determined by the dependence

$$\sigma_{zTf} = \frac{(T_m^0 - T_p^0)(\alpha_s - \alpha_f)}{\delta_f \Psi k^2} \cdot \frac{\sinh kb - \sinh kx - \sinh k(b-x)}{\sinh kb} \tag{4.3}$$

Normal stresses appearing in the flange are of the opposite sign, and their value can be defined by substituting shoulder thickness δ_s for flange thickness δ_f in Eq. (4.2).

Because of the difference in Poisson ratios at tension and compression, edge effects also arise along the z axis. The value of tangential stresses can be determined by the relation

$$\tau_{zy\mu} = \frac{A}{\Psi k} \frac{\cosh kz - \cosh k(b-z)}{\sin kb} \tag{4.4}$$

where

$$A = \mu_{xzs} \frac{N_s - N_{Ts}}{\delta_s E_{xs}} - \mu_{xzf} \frac{N_{Tf}}{\delta_f E_{xf}} \tag{4.5}$$

N_s, N_{Ts}, and N_{Tf} are linear stresses in the shoulders and flange exerted by external force factors and temperature; E_{xs}, E_{xf}, μ_{xzs}, and μ_{xzf} are the elasticity module and Poisson ratios of the shoulders and flange in the axial direction.

Normal stresses along the z axis are calculated by the formula

$$\sigma_{zf} = \frac{A}{\delta_f \Psi k^2} \cdot \frac{\sinh kb - \sinh kz - \sinh k(b-z)}{shkb} \tag{4.6}$$

Tangential stresses operating in combination on free edges of the flanges are the reason for the edge effects that initiate breakage.[262]

The methods of fighting this phenomenon include design debugging with strengthening free edges through stitching or curling the outer layer and packing either a winding or a braiding layer along the article perimeter, as well as performing other operations.

Experience in application of such complex-profile structures as well as experimental investigations have proved that the origin of breakage and, consequently, the weakest area of the structure, are in the profile center. As the results of works in Refs. 45, 121, and 180 have revealed, the causes are the main normal stresses. In this connection, it is recommended that flanges be additionally stiffened with layers of fibrous reinforcement at a 90° angle to the rod axis, which is most adept at perceiving such stresses.

Nevertheless, a more comprehensive analysis of failure behavior reveals that the model used[262] is not quite adequate. It has been established that failure most often starts as exfoliation at the shoulders in the negative-curvature zone of the profile, as an inner interlayer breaks or tapers out of the flange elements together with the shoulders near the negative-curvature zones.[141] This failure may occur under tensile and compression stresses. Apart from this, delaminating of the profiles is observed during their processing or cooling after high-temperature polymerization.[54,96]

Failure can occur both from tensile and from compressing stresses.

Note that the integrated braiding–pultrusion procedure as applied to complex-profile articles may result in variation of the composite structure and, consequently, in altered stress-strain state character of their elements.

4.2 Stress–Strain State Analysis of Complex-Profile Rod Elements

For analyzing the stress–strain state, it is necessary, first, to specify the design model of the profile article, clarify the causes of early failure, and outline the required structural and technological refinements. In addition, allowances should be made of the structural peculiarities that are to be exhibited at the braiding–pultrusion procedure.

As an example, let us take a double-T shaped profile as one of the most applicable and characteristic of all complex-profile product types.

To improve the service characteristics of the post and stiffen its mating zone with the flange, both the post and its mating zone can be given additional reinforcement, which makes the flange thickness variable. This is highly important for the braiding–pultrusion technique because a thick sleeving layer on the article periphery is able to decrease its bearing capacity at bending whereas a thin layer does not add reliability to the post–flange junction.

As a result, we obtain a design such as that presented in Fig. 4.3*a*. The primary flange material is a unidirectional composite with an isotropic plane and can be treated as an isotropic material when solving the plane problem.

Apparently, the primary flange material of unidirectional composite has an isotropic plane and at the solution of a flat problem of theory of elasticity can be watched as an isotropic material.

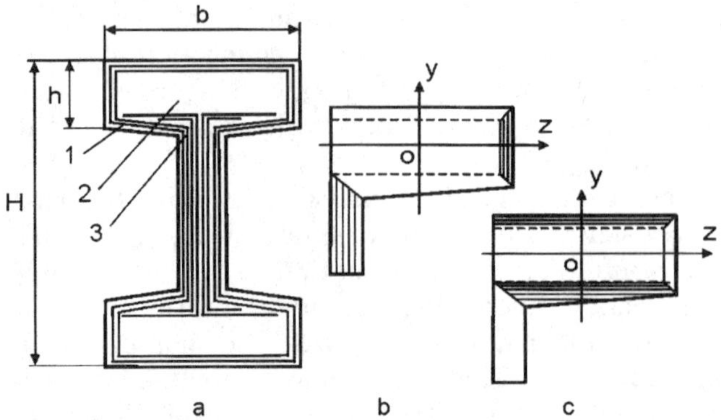

Figure 4.3 Structure and design models of a profile element: (*a*) profile design, (*b*, *c*) face and side models of the flange portions; (1) braid; (2) main reinforcement; (3) insertions.

Similarly to the braiding material, the insertions display different characteristics in longitudinal and transverse directions, the longitudinal direction vector being always oriented over the article perimeter.

Physic-mechanical characteristics of the braid in the transverse direction and the properties of the main reinforcement of the flange in cross-sectional plane are very close and can be considered identical.

The design model of the flange face portion can then be taken to be of the form presented in Fig. 4.3*b*, and that of side portions, as in Fig. 4.3*c*.

In Fig. 4.3, along with the constant-thickness portions, areas of variable thickness where the layers are glued together are also observed. The material of the braid and insertions in the bending point are presented as two structural elements made of materials with different properties and glued to one another at a 45° angle.

Because of varying temperature conditions and differences in Poisson ratios of the main material of the flange and braid, one can observe tangential at the interfaces and normal in the actual layer stresses. These stresses were found to affect the article's bearing capacity.

To detect critical points in the cross section, as well as the type and magnitude of the stresses acting in them, let us consider an element of the adhesive joint of length dz that incorporates both flange and braid layers (Fig. 4.4).

To consider a single adhesive layer when calculating τ_{zy}, the layer thickness δ_2 can be taken equal to $\delta_s/2$ because tangential stresses τ_{zy} along the axis should be equal to zero. Then, when τ_{yz} is calculated, the layer thickness δ_2 will be $b_s/4$ due to analogous properties of τ_{zy}.

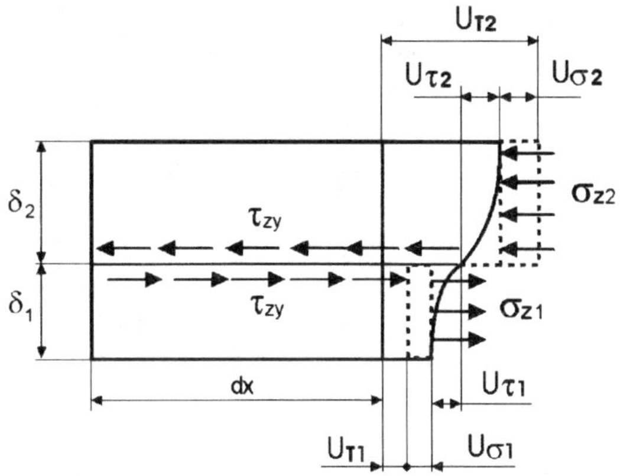

Figure 4.4 Design model of an adhesive layer at thermal effects.

The continuous function τ_{ij} changes its sign at the indicated points. Generally, it is taken that the thickness of both layers along the y axis is variable and that it is a function of z.

To simplify the calculations, we shall take into account only thermal stresses. When analyzing Poisson stresses, the reasoning will be the same, except for the causes of deformations.

In conditions of free deformation in varying temperatures, each element would change dimensions by a value $u_{Ti} = \alpha_i \Delta T dz$. Because the elements are fastened together and deform jointly, then in a general case, both normal and tangential stresses appear in them. So, displacements in the splice plane are found as follows:

$$
\begin{aligned}
u_1^* &= \alpha_1 \Delta T dz + u_{\sigma 1} + u_{\tau 1} \\
u_2^* &= \alpha_2 \Delta T dz + u_{\sigma 2} + u_{\tau 2}
\end{aligned}
\tag{4.7}
$$

where α_i is the LTEF of the layers and $u_{\sigma i}$ and u_{Ti} are displacements in each layer in response to normal and tangential stresses.

Bearing in mind the above stipulated zero conditions, and taking the first approximation of the linear distribution of tangential stresses across each layer thickness, the integral shift of the splice plane can be defined by the formula

$$
u_{\tau i} = \frac{\tau_{zy}\delta_i}{2G_i} \tag{4.8}
$$

With respect to joint deformation, the left-hand and, consequently, right-hand, sides of Eq. (4.5) are equal to each other.

Let us differentiate the obtained equation along z, considering that $u_{\sigma i}$, τ_{zy}, and δ_i are the functions of this variable

$$(\alpha_1 - \alpha_2)\Delta T + \frac{du_{\sigma 1}}{dz} + \frac{du_{\sigma 2}}{dz} + \left(\frac{\delta_1}{2G_1} + \frac{\delta_2}{2G_2}\right)\frac{d\tau_{zy}}{dz} + \left(\frac{d\delta_1}{2G_1\,dz} + \frac{d\delta_2}{2G_2\,dz}\right)\tau_{zy} = 0 \tag{4.9}$$

Thus,

$$\frac{du_{\sigma i}}{dz} = \varepsilon_{\sigma i} = \frac{\sigma_i}{E_i} \tag{4.10}$$

Let us differentiate repeatedly along z and come to $d\sigma_i/dz$ from the differential equilibrium equation

$$\frac{d\sigma_i}{dz} + \frac{d\tau_{zy}}{dy} = 0 \tag{4.11}$$

Taking the assumption of a uniform distribution of tangential stresses over area dz and normal stresses across the element thicknesses, we then obtain

$$\frac{d\sigma_i}{dz} = -\frac{\tau_{zy}}{\delta_i} \tag{4.12}$$

Upon substitutions and division of all members of the equation by the second-derivative factor, we obtain a homogeneous second-order differential equation with respect to τ_{zy}

$$\frac{d^2\tau_{zy}}{dz^2} + 2\eta(z)\frac{d\tau_{zy}}{dz} + k^2(z)\tau_{zy} = 0 \tag{4.13}$$

where

$$\begin{aligned} \eta &= \frac{1}{2\Psi}\left(\frac{d\delta_1}{dz} + \frac{d\delta_2}{dz}\right) \\ k^2 &= \frac{1}{\Psi}\left(\frac{1}{E_1\delta_1} + \frac{1}{E_2\delta_2} - \frac{d^2\delta_1}{2G_1\,dz^2} - \frac{d^2\delta_2}{2G_2\,dz^2}\right) \\ \Psi &= \frac{\delta_1}{2G_1} + \frac{\delta_2}{2G_2} \end{aligned} \tag{4.14}$$

Thus, Eq. (4.13) is a relation with variable factors that cannot be solved with elementary functions. To solve it, standard numerical software should be used, e.g. Mathcad-2000. For boundary conditions, normal stresses at extreme points of the profile and displacements in the

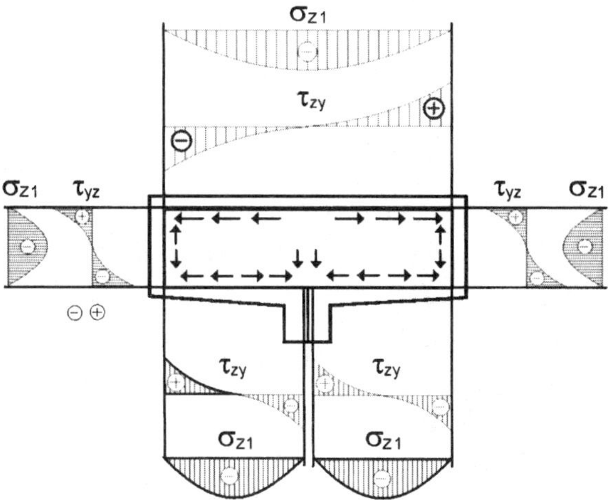

Figure 4.5 Stress distribution in the braided layer over the profile flange contour.

profile symmetry axis are taken to be equal to zero. In addition, as was specified earlier, tangential stresses in the points coinciding with coordinate axes are taken to be of zero order.

When element thickness values are constant, δ_i = constant, $\eta = 0$, Ψ = constant, and k^2 = constant. The equation is then reduced to known solutions[180,262] in elementary functions, so the stresses can be calculated by Eqs. (4.1) to (4.6).

The qualitative character of distribution diagrams of normal and tangential stresses over the profile contour is illustrated in Fig. 4.5.

Calculation results using the model described have revealed two dangerous points in similar profile articles: one on the free edge, as in the previous model, and the other inside the structure. We emphasize that the tangential stresses inside each shoulder are lower because of the variable stiffness of the joint.[45] If the shoulder is thin enough, this value is insignificant, so tangential stresses remain hazardous.

As follows from analysis of dependencies, the maximum normal and tangential stresses are not, in fact, dependent upon the lengths of the interacting elements. However, these values are closely interrelated with the material characteristics of these elements and their thickness.

Variation curves of the functions of the maximal normal and tangential stresses due to thermal effects are shown in Figs. 4.6 and 4.7 as being dependent on the relative thickness of the braiding layer and its elasticity modulus. Calculations were made for a complex-profile carbon plastic rod element with temperature variation $\Delta T = 200°$C. As the diagrams show, the maximal normal and tangential stresses with elasticity modulus augment the growth of the braiding layer.

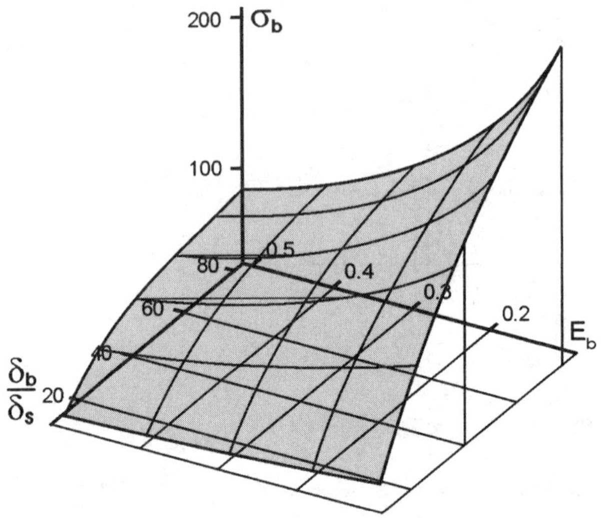

Figure 4.6 Behavior of normal thermal stresses in the braided layer with changes in thickness and elasticity.

As for the braid thickness $2\delta_b/\delta_s$, the maximum normal stresses diminish with increasing thickness in contrast to tangential stresses, which increase.

A peculiarity of the present model is different signs of tangential stresses in the left and right shoulders that are summed up in the

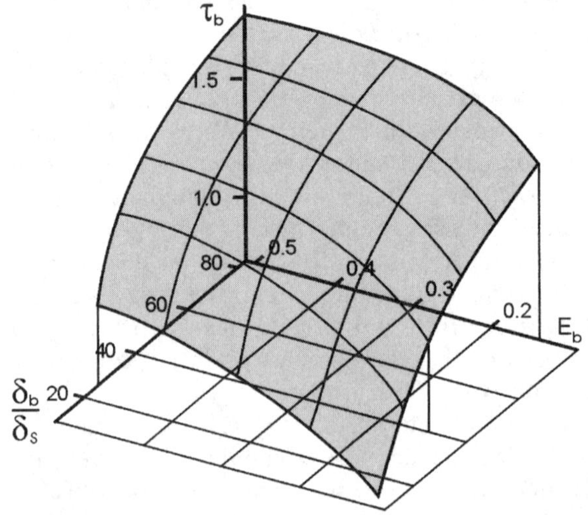

Figure 4.7 Behavior of maximum tangential thermal stresses in the braided layer with changes in thickness and elasticity.

direction of the Y axis. In addition, tangential stresses induced by outer loads are added to them, and their combined action is causes the observed failure mechanism. So far, the internal zone of profile articles of a double-T type turns to be more critical than the edges with respect to stress concentration and feasible breakage onset.

As recommended,[180,252] introduction of additional reinforcement in the direction of the z axis changes physic-mechanical parameters of the flange, bringing them closer to those of the shoulders and, consequently, assisting in reduction of stresses and structural strengthening. The design model in this case will undoubtedly correspond to the one proposed.[262]

It is reported[45,121] that addition of $\pm 45^{\circ}$ layers into the shoulder design has impaired longitudinal strength and stiffness. The transverse reinforcing layers are commonly not introduced to avert further reduction of strength and stiffness of the shoulders in the longitudinal direction. It is evident that strengthening should be reached by some other methods proceeding from the analysis of a refined model and failure concept, e.g. by stitching the shoulders. As a result, normal stresses are transmitted by the stitching material while tangential stresses in the central part of the article tend to zero.

As shown,[65,152,289,290,335] high serviceability of unidirectional composites at compression can be attained only in the presence of a closed convex supporting shell that hinders fibers from losing stability and from breaking of the rods via delaminating.

It is recommended that the closed shell be produced by stitching the braiding layer. If, to follow the braiding–pultrusion production method described for rod articles, the tangential stresses at angle points of the shelves will occur but their magnitude will be cut in half, as the stresses are to be propagated over two zones of the adhesive contact. Normal stresses are distributed more evenly over the main reinforcement of the shelves, so the system appears to be more in equilibrium in mechanical and thermal respects.

4.3 Calculations and Choice of Structural Parameters of Complex-Profile Rod Elements

The basic structural parameters of complex-profile rod elements intended for calculations are the following:

1. The type of the fibrous reinforcement for flanges experiences tensile and compressive loads exerted by axial and bending moments during operation.
2. The composite material used to form posts perceives mainly tangential stresses from transversal forces at bending.

3. The type of the fibrous reinforcement intended for the braiding layer bears tension–compression forces that are generated in profile flanges as a result of temperature and Poisson strains.
4. The type of the fibrous reinforcement used at profile stitching and closure of the flange braiding layer, which also perceives the loads specified above.
5. Thickness and width of the flange basic material.
6. Thickness of the post and shoulders.
7. Packing angle and braiding layer thickness.

4.3.1 Selection of the complex-profile article material

The type of the fibrous reinforcement for molding principal bearing elements of the profile article is chosen considering its designation, operating loading regimes, material properties, and a series of economical and technological factors.

For aviation and space vehicles, carbon-based, boron, and organic fibers are most often employed for their high specific strength and stiffness indices. It is preferable that products operating under strong compressive stresses have thicker boron fibers, which are more resistant to loss of stability. In contrast to boron fibers, which are rather expensive and difficult to process, organic plastics can ensure high strength at tension and are sufficiently cheap, although they display poor resistance to compression.

Considered to be most promising are carbon fibers, as they can guarantee the best correlations of various, often contradictory requirements. Depending on the target, they are subdivided into high-modular and high-strength fibers. High-modular fibers are employed in structures whose paramount requirement is to ensure the needed stiffness, and high-strength fibers are employed where strength is most important.

The types of fibrous fillers described above are rather costly, and their application in other fields is restricted, although there is a stable tendency toward cost depreciation as their consumption grows.

The most applicable fillers today have turned out to be glass, which are the cheapest and most practically feasible at processing. The problems of selecting composite materials have been set forth,[45,87,99,121,288] their applications in different domain are cited,[225,257] and the calculation methods have been discussed.[103,111,115,116,182,219,301]

Investigations into the physic-mechanical characteristics of the composites used in rod structures are reported,[25,155,243,271,304,348] and their behavior under various loads is explained.[48,72,122,263,274,314,427,428]

The questions of manufacturing materials by pultrusion are answered in various works,[30,34,42,44,49,88–90,94,101,259,260,271,361] and questions of manufacturing by braiding are discussed by many authors.[74,78,146,169,187,255,338,419] Numerous works are devoted to studying knitted materials[130,286,304,308,386,401] as well as woven materials.[57,59,148,168,175,223,375,418]

In conclusion, it is expedient to use woven strips at the molding post of profile ware with a fiber packing angle of $\pm 45^\circ$. The strip material is usually chosen to be the same as the material used for the main reinforcement of the profile flange, although hybridization is now being employed more often to cut costs and improve product characteristics. Fabrics that are commonly used are based on organic, glass, or carbon fibers.

The braiding layer is formed of interwoven braids of the fibrous filler. The choice of the braiding material is similar to that of the post, and they need not match.

In principle, the profile can be pierced by different methods, including composite or metal short rods, clips, threads, and fibrous braids, depending on the article dimensions and required mechanical characteristics of the stitching layer.

Optimum conditions for rational reinforcing of composite structures by short-fiber particles have been formulated elsewhere.[354] It is emphasized in that work that, with respect to quality, a positive effect of radial reinforcement and stitching is exhibited; in case a series of regularities are obeyed, the following points should be considered:

1. Fiber diameter should be commensurable with that of the main reinforcement and must not induce damage or distortion.
2. Short fibers are to be sufficiently rigid and strong.
3. Adhesive bonds of the main reinforcement should be equivalent to stitching fibers.

Apart from these points above, an additional important requirement of the stitching material remains: it must be undamaged during piercing under friction and dynamic loads. To achieve this requirement, lisle glass threads and organic threads or braids, or carbon braids with high textile characteristics are employed.

4.3.2 Computation of the main flange material

When estimating the primary flange material of a complex-profile article, it is expedient to make use of investigation results cited[30,202] and recommendations presented.[88,120,121]

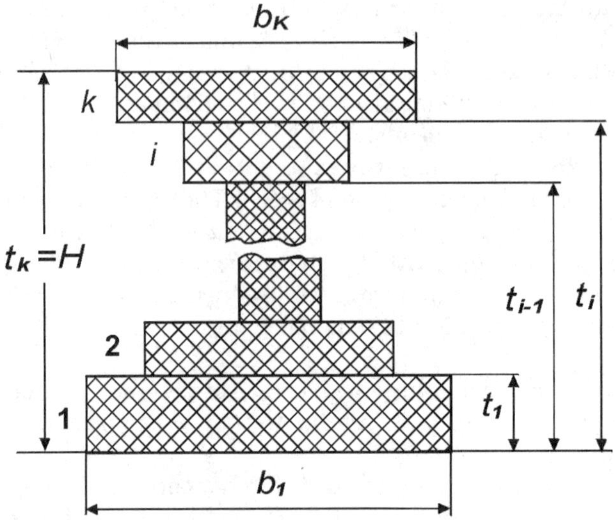

Figure 4.8 Design model of a complex-profile article.

A beam is calculated as a laminated structure consisting of k layers of thickness δ_i and width b_i, where each layer is at a distance $t_{i-1} = \sum_{j=1}^{i-1} \delta_i$ from the origin (Fig. 4.8).

The displacement distribution in the section conforms to the hypothesis on plane sections that is widely applied in calculations of the beams. Thus far, normal stresses in each point of a section along the beam length can be found from

$$\sigma_{xi} = \frac{E_{xi} M_z}{D} \bar{y}_i \tag{4.15}$$

where E_{xi} = elasticity modulus of the ith layer in the x axis direction
$\bar{y}_i = y - e$ = distance from the neutral axis to the point under study of the ith layer
D = generalized stiffness of the beam cross section at bending

$$D = I_2 - \frac{I_1^2}{I_0}, \qquad e = \frac{I_1}{I_0} \tag{4.16}$$

$$I_n = \frac{1}{n+1} \sum_{i=1}^{k} b_i E_{xi} \left(t_i^{n+1} - t_{i-1}^{n+1}\right) \quad n = 0, 1, 2 \tag{4.17}$$

The needed bearing capacity of the profile article will then be reached by obeying strength conditions for the outside fibers of the main material of the flange.

For symmetry of the section about the z axis

$$\sigma_{xs}^{\max} = \frac{E_{xs} M_z H^*}{2D} \leq [\sigma]_{\min} \tag{4.18}$$

The required value of the generalized stiffness is

$$D \geq \frac{E_{xs} M_z H^*}{2[\sigma]_{\min}} \tag{4.19}$$

where $[\sigma]_{\min}$ is the least of admissible tensile and compressive stresses and $H^* = H - 2\delta_{\mathrm{b}}$ is the profile height without outer braiding layers.

To design a simplified calculation model of the generalized stiffness it is acceptable to make an allowance for the main material of the flanges only

$$D = b_{\mathrm{s}} E_{xn} \left(\frac{2}{3}\delta_{\mathrm{s}}^3 + \frac{H^{*2}\delta_{\mathrm{s}}}{2} - H^* \delta_{\mathrm{s}}^2 \right) \tag{4.20}$$

We then obtain a cubic equation relative to δ_{s} by substituting Eq. (4.19) in Eq. (4.20) and arrive at an analytical solution

$$\frac{2}{3}\delta_{\mathrm{s}}^3 - H^* \delta_{\mathrm{s}}^2 + \frac{H^{*2}}{2}\delta_{\mathrm{s}} - \frac{M_z H^*}{2b_{\mathrm{s}}[\sigma]_{\min}} = 0 \tag{4.21}$$

Nevertheless, in practice the profiles usually meet the condition $\delta_{\mathrm{s}} < H^*/4$.

Therefore, the cubic member Eq. (4.21) can be neglected, and δ_{s} is then found to an accuracy of 5 percent from the quadratic equation

$$\delta_{\mathrm{s}}^2 - \frac{H^*}{2}\delta_{\mathrm{s}} + \frac{M_z}{2b_{\mathrm{s}}[\sigma]_{\min}} = 0 \tag{4.22}$$

$$\delta_{\mathrm{s}} = \frac{H^*}{4} - \sqrt{\frac{(H^*)^2}{16} - \frac{M_z}{2b_{\mathrm{s}}[\sigma]_{\min}}} \tag{4.23}$$

The "+" sign contradicts the accepted assumption so that the solution obtained is unambiguous.

Prior to using the solution of Eq. (4.23), one can define the width b_{s} from the condition of the determinant positivity

$$\frac{(H^*)^2}{16} - \frac{M_z}{2b_{\mathrm{s}}[\sigma]_{\min}} \geq 0$$

where

$$b_{\mathrm{s}} \geq \frac{8M_z}{(H^*)^2[\sigma]_{\min}} \tag{4.24}$$

We may set a tentative height-to-width ratio of the profile article $H^*/B = \alpha$ and calculate the profile height

$$H^* \geq \sqrt[3]{\frac{8M_z\alpha}{[\sigma]_{\min}}} \tag{4.25}$$

4.3.3 Calculation of post thickness

Tangential stresses in the profile post can be found from the relation

$$\tau_{xy} = -\frac{Q}{Db_r}\int_0^y E_{xi}b_i\bar{y}\,dy \tag{4.26}$$

and their maximum values can be substituted into the strength condition.

For a symmetrical profile,

$$\tau_{xy}^{\max} = \frac{Q}{Db_r}\int_0^{\frac{H}{2}} E_{xi}b_i\bar{y}\,dy \leq [\tau] \tag{4.27}$$

Due to the small thickness of the braiding layer, a stress increment in it can be neglected

$$\frac{Q}{b_r D}\left(E_{xr}b_r\int_0^{\frac{H^*}{2}-\delta_s} y\,dy + E_{xs}b_s\int_{\frac{H^*}{2}-\delta_s}^{\frac{H^*}{2}} y\,dy\right) \leq [\tau]$$

The desired post width of the profile can then be calculated by the formula

$$b_r \geq \frac{E_{xs}b_s\delta_s(H^* - \delta_s)}{\frac{2[\tau]D}{Q} - E_{xr}\left(\frac{H^*}{2} - \delta_s\right)^2} \tag{4.28}$$

D can be found from Eq. (4.20).

4.3.4 Calculation of braiding layer parameters

The chief loads perceived by the braiding layer are the normal stresses acting in the z direction. These stresses arise during interactions between the braiding layer, shoulders, and the main material of the flange because of differences in their physic-mechanical characteristics. They reach their maximum in the braid on the profile periphery in the flange center loaded by the axial tensile stresses. Just here, Poisson and thermal stresses sum up because their vector directions coincide. The braiding layer is simultaneously loaded by compressive stresses.

When the braiding layer thickness and that of the main flange material are constant in this area, stresses can be calculated by Eqs. (4.3) and (4.6), and the strength condition of the braided layer will be written as follows:

$$\sigma_{zb}^{\max} = \frac{(\alpha_{zs} - \alpha_{zb})\Delta T^0 + A_H}{\delta_b \Psi k^2} \leq [\sigma_b^-] \tag{4.29}$$

Thermal constituents of longitudinal strains are rather feeble, and, as follows from (4.5), they reduce the value A_H. Hence, they can be neglected at designing if they contribute to the strength margin. Thus,

$$A_H = \mu_{xzs} \frac{2N_s}{\delta_s E_{xs}} \tag{4.30}$$

where

$$N_s = \int_{\frac{H^* - \delta_s}{2}}^{\frac{H^*}{2}} \sigma_x \, dy$$

We substitute σ_x in Eq. (4.15) and obtain

$$A_H = \mu_{xzs} \frac{M_z(2H^* - \delta_s)}{4D} \tag{4.31}$$

The product of Ψk^2 is calculated by the relation

$$\Psi k^2 = \frac{2}{E_s \delta_s} + \frac{1}{E_b \delta_b} \tag{4.32}$$

It is evident that the braiding layer thickness δ_b is included in Eq. (4.32).

By substituting values of Eqs. (4.31) and (4.32) into the strength condition (4.29) and by certain manipulations, we obtain the following formula for the braiding layer thickness:

$$\delta_b \geq \frac{E_s \delta_s}{2} \cdot \left[\frac{(\alpha_{zs} - \alpha_{zb})\Delta T^0 + \mu_{xzs} \frac{M_z(2H^* - \delta_s)}{4D}}{[\sigma_b^-]} - \frac{1}{E_b} \right] \tag{4.33}$$

The Poisson and thermal stresses in the flange loaded by compressive strains have opposite signs, so the braid material is less loaded. Thermal stresses arise in the course of temperature treatment of the material and relax with time. Hence, the primary stiffening function the braiding exercises at flange loading is only by force factors, in our case by M_z. As for the braiding material, it is loaded by tensile stresses, so strength limit $[\sigma_b^+]$ should be substituted into Eq. (4.33).

Analysis of the relation proves that, in case of low elastic characteristics of the braiding layer, a situation may arise when the right-hand side of the equation will be negative. As a result, the braid ceases to stiffen the structure but operates simply as one of the flange materials. Proceeding from this, definite recommendations can be worked out on the choice of the braiding material and its parameters.

To ensure the effectiveness of the stiffening function, the following condition should be met:

$$\mu_{xzs}\frac{M_z(2H^* - \delta_s)}{4D[\sigma_b^+]} - \frac{1}{E_b} > 0 \tag{4.34}$$

It then follows that the braiding material must conform to the requirement

$$\frac{[\sigma_b^+]}{E_b} < \mu_{xzs}\frac{M_z(2H^* - \delta_s)}{4D} \tag{4.35}$$

since the relation $[\sigma_b^+]/E_b$ is a function of the reinforcement angle of the braiding material; Eq. (4.35) may then be used for the smallest value at which braiding will acquire stiffening properties.

4.3.5 Calculation of stitching parameters

The major loads experienced by the stitching material are the maximum normal stresses operating in the flange junction with the post. Stiffness of the stitching elements is by two orders of magnitude higher than that of the composite being pierced in transversal direction and just these elements perceive all loads.

The linear load value operating in the lower part of the flange in the stitching zone can be calculated as follows:

$$P = \frac{A_B}{\Psi k^2} \tag{4.36}$$

The value A_B for the inner flange part can be found from Eq. (4.30), where the longitudinal force value N_s is determined by

$$N_s = \int_{\frac{H^*}{2}-\delta_s}^{\frac{H^*-\delta_s}{2}} \sigma_x \, dy$$

Thus,

$$A_B = \mu_{xzs}\frac{M_z(2H^* - 3\delta_s)}{4D} \tag{4.37}$$

The strength condition for stitching elements can be written as

$$\sigma_W = \frac{P}{F_W n_W} \leq [\sigma_W] \tag{4.38}$$

where $n_W = 1/t_W$ is the number of stitching elements per meter and t_W is the seaming pitch (the distance between neighboring stitching elements).

When threads or braids of fibrous materials with a known linear density T are used for stitching, the total area of the fibers can then be expressed through

$$F_W = \frac{T}{\gamma_W} \cdot 10^{-6} (m^2) \tag{4.39}$$

where γ_W is the specific weight of threads and braids (in kg/m^3).

The design formula for determining seaming pitch can be obtained upon substitution of all values in the strength condition (4.38)

$$t_W \leq \frac{4D\,\Psi k^2 T[\sigma_W] \cdot 10^{-3}}{\mu_{xzS} M_z (2H^* - 3\delta_S)\gamma_W} \text{ (mm)} \tag{4.40}$$

Using this relation, one can select optimum correlations of the stitching material, linear density of the threads or braids, and seam pitch.

4.4 Conclusions

1. Existing design models for analysis of the stress state of profile structures do not always adequately describe the behavior of real objects. In this connection, a design model is proposed, which is able to interpret certain experimental observations and failure effects that occur at operation.
2. Analysis of the results obtained shows that both tangential and normal stresses in their concentration points and normal stresses in their maximums may pose dangers for the structures. Their magnitude in articles produced by the braiding–pultrusion method is half the magnitude as the products devoid of a closed braiding layer.
3. Tangential stresses of the left and right parts of the flange sum up in the junction zone of the flange and post, and the material is then in a hazardous strained state. The stress concentration can be corrected for by introduction of an additional layer having physic-mechanical characteristics that approach those of the shoulders. It is also possible to fix the braiding layers mechanically by stitching.
4. Investigations described herein have revealed the principal structural parameters of the typical integral structure of a complex-profile rod element and have formed the basis for their design procedures.

Chapter

5

Analysis of Force Parameters

A critical parameter of the braiding–pultrusion process is the magnitude of the pulling force. It increases until it reaches a threshold that may lead to rupture of the longitudinal reinforcement, which would hinder further molding.[105,114,240] An optimum level and certain limiting values of the drawing force are essentially dependent on the cross-sectional area of the profile molded, its geometrical parameters, and filling degree along with the strength of braids or fibers that contribute to the total strength of the reinforcing bunch of fibers. The drawing force is thus an integral parameter of the process. The factors mentioned above together with thermal and kinetic conditions of the molding process exert determining the quality of articles produced.[288] Thus far, it has been necessary to regulate the drawing force, which can be diminished during the process by alleviating the tension force of the braiding threads, that of the main reinforcement, and the rated dimensions of the article formed and by lowering the binder temperature at different process stages. This will then lead to changes in the article stress–strain state, avert warping, and cut energy consumption.[119,209,298]

5.1 General Characteristics of the Process

The magnitude of the pulling force is a combination of forces that arise from the resistance to motion of the article being formed at individual stages of its production. Initially, reinforcing fillers leaving the reels prior to the impregnation bath are stretched with a definite tension to make the longitudinal reinforcement linear at all stages. This force value, being roughly 10–15 N/braid, is chosen experimentally and is

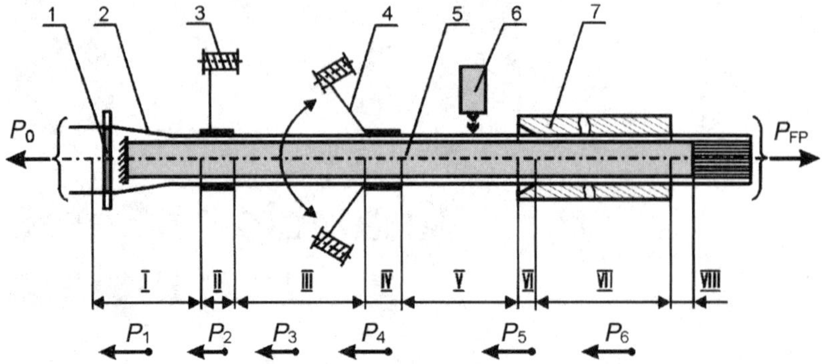

Figure 5.1 Scheme of distribution of forces affecting the article being pulled along the impregnation-molding channel: (1) insulating comb; (2) longitudinal reinforcement braids; (3) transversal reinforcement braids; (4) wound braids of the reinforcement layer; (5) mandrel; (6) piercing unit; (7) forming die.

maintained by idle rollers of the strainer through which the multiple folded braids are passed. As they arrive at the impregnation bath, the braids are wetted in preheated binder, and are slowed down, thanks to their folds, to achieve an additional tension force. The total strain at the bath outlet is insignificant, up to 20–25 N/braid.

Later in the process, the saturated braid is fed into the principal molding unit. This motion can be presented as a combination of separate stages during which additional constituents of the pulling force appear. The distribution of these forces for a combined process that consists of the packing of the longitudinal reinforcement, cross winding, braiding, piercing, and molding of the article in a die is illustrated in Fig. 5.1.

Zone 1 is characterized by the motion of a rectilinear braid impregnated with a binder over the mandrel surface. The resultant resistance force P_1 is governed by the binder and mandrel cohesion forces. These cohesion forces can be found experimentally when the ambient temperature, binder viscosity, and contact area are taken into account. In zone II, the material cross-section is formed in case a transversal reinforcement within the structure is needed. This is made by cross-winding of a group of threads with a certain tension. As a result, a definite pressure is exerted on the material packet over some length from the side of the wound layer. By determining this pressure value and knowing the sliding coefficient, we can determine the resistance force P_2. Concerning the tension force of the winding thread, it should be able to provide for the required dimensions of the cross-layer during winding, then, because of the binder squeezing at the end of this zone, the pressure within the packet structure tends to zero. The motion in zone III can then be characterized simply by sliding of the formed structure over the mandrel surface. The resistance force P_3 in this case is defined by the impregnated material cohesion with the mandrel. In zone IV, oblique cross-threads are laid on the material surface, which is called braiding. As in zone II, the binder is filtered through the braiding layer as a result of outside reduction, making the pressure at the end of this zone in the material packet equal to zero. One peculiarity of such a structure formation within this zone is the tension of the winding reinforcement at a certain angle to the mandrel axis; as a result, an additional resisting force P_4 arises from a sum of projections of tension forces of each thread along the mandrel axis. Motion in zone V is analogous to that in zone III, and the arising resistance force P_5 (EMBED Eq. 5.3) is determined by the same parameters.

The die that is chosen molds articles of a needed size and performs their thermal treatment. Three main parts can be conventionally singled out in a die. In the first of them (zone IV), the article cross-section is molded, during which the structure is reduced and excess binder is squeezed out and removed. The resultant backpressure hampers displacement of the article. In the following parts of the die, the binder is cured under certain temperatures, which is accompanied by the formation of a honeycomb polymer and conversion of the binder from a viscous state to a solid body. During solidification at a constant temperature, either the initial viscosity of the binder or its rigidity at the final stage may rise abruptly as a result of both gel formation and its glass transition due to augmented molecular weight, formation of new chemical and physical bonds, and incremented cellular density. The chief characteristics of solidification are considered to be volume shrinkage and the amount of low-molecular volatiles liberated. Volume shrinkage during solidification in zones IV and VIII is brought about by an increased density of the binder at setting because of the great number of chemical bonds formed. As a result of shrinkage, the pressure transferred onto the inner mandrel rises and the resistance forces to motion increase because of intensified friction. However, a question remains on the variation of the friction coefficient with article motion from the die to the outlet.

Analysis of the process layout has emphasized that the total pulling force depends upon a series of interrelated parameters, a part of which can be preset within the technological process proceeding from the condition of diminishing resistance to pulling. The other part is specified by the character of chemical and physicochemical reactions taking place at molding. On the whole, the total pulling force is the result of resistance forces operating at each stage of the process.

$$P_{\mathrm{FP}} = \sum_{i=0}^{i=8} P_{\mathrm{i}} \tag{5.1}$$

5.2 Determination of Pulling Resistance at the Cross-Winding Stage

The principal technological parameters of this stage on which the required structure of the material is dependent are the throughput rate v_S, tension force of the winding reinforcement N_0, winding velocity n_0, binder viscosity η, and geometrical dimensions of the winding reinforcement. Note that winding velocity, throughput rate, and winding reinforcement width are closely interrelated

$$V = t \cdot n_0 \tag{5.2}$$

where t is the layout pitch equal to the winding reinforcement width.

During winding, the amount of the binder is redistributed, i.e., a viscous flow occurs that can be described by the filtration theory of liquids in porous media.[50,55,97,194,216] In this case, the binder is commonly presented as a viscous, incompressible liquid obeying Darcy's law at filtration, where the reinforcement stress state is neglected.

Under the action of the tension force of the winding reinforcement, there appears a radial constituent of the normal pressure that molds the longitudinal reinforcement

$$T = \frac{2N_0}{d_0\sqrt{1 + tg^2\gamma}} \tag{5.3}$$

where γ is the winding angle and d_0 is the outer diameter of the article inner layer.

By setting the pressure distribution to be uniform on the molding, and lengthwise along the article being wound to a distance equal to the winding pitch, we obtain a relation for pressure

$$T = \frac{2N_0}{d_0 t\sqrt{1 + tg^2\gamma}} \tag{5.4}$$

Redistribution of the binder starts as soon as it is laid on the winding reinforcement of the longitudinal layer. The flow of the viscous, incompressible liquid at constant temperature is described by the Navier–Stokes equation,[349] which is presented in vector form as

$$\frac{d\vec{V}}{d\tau} = \vec{F} - \frac{1}{V}\vec{\nabla}T - \frac{\eta}{V}\nabla^2\vec{V} \tag{5.5}$$

In supposition of the laminar flow of the liquid binder and neglecting gravity, we obtain Carman–Kozeny's equation reduced in the theory of filtration to the form[201]

$$\vec{V} = \frac{K_p T}{\eta L_0} \tag{5.6}$$

In this case, $\vec{V}$ is the mean velocity of the liquid phase flow, K_p is the permeability ratio dependent upon the specific area of particles (S_V), porosity, and tortuosity of the channels L_Q

$$K_p = \frac{1}{K_6}\left(\frac{L}{L_Q}\right)^2 \frac{U^2}{S_V^2} \tag{5.7}$$

where U = volume fraction of the liquid phase
S_V = specific area of the solid phase
K_6 = factor found in Refs. 118, 126, 201, and 216

Under the effect of molding T squeezed from the considered volume binder moves in two directions: along the article axis opposite to velocity vector V_S and in a radial direction moving to the outer surface and, if necessary, saturating the winding reinforcement. The total amount of the binder that penetrates through the surfaces considered will be

$$Q = Q_L + Q_r \tag{5.8}$$

where Q_L is the amount of the binder filtrated in the longitudinal direction and Q_r is that in the radial direction.

For the known initial filling degree φ_B and the final φ_E, required one, the amount of binder removed at winding will then be

$$Q = \frac{\pi\left(d_0^2 - d^2\right)\cdot t\cdot(\varphi_E - \varphi_B)}{4\varphi_E} \tag{5.9}$$

where d is the inner diameter of the mandrel.

Thus, the amount of the binder that moves in axial direction with time and corresponding to the winding pitch is

$$Q_L = \frac{\pi\left(d_0^2 - d^2\right)}{4} - (1-\varphi_B)(\vec{V} - V_S)\frac{t}{V_S} \tag{5.10}$$

Let us substitute instead of $\vec{V}$ its value from Eq. (5.6) in the given equation with account of Eq. (5.7). Presuming that the fibers are slightly tortuous, such that $L/L_Q = 1$, $U = 1 - \varphi_B$, and $S_V = 4\varphi_B/d_f$, we then obtain

$$Q_L = \frac{\pi\left(d_0^2 - d^2\right)(1-\varphi_B)t}{4}\left[\frac{(1-\varphi_B)^2 d_f^2\, T_0}{16\eta\, L_0 K_6\, \varphi_B^2\, V_S} - 1\right] \tag{5.11}$$

where d_f is the fiber diameter and L_0 is the channel length that can be taken as equal to laying pitch t at a given flow form.

It is also possible to determine the amount of the binder moving in the radial direction, i.e., crosswise to the fiber arrangement in the structure where the distance between fibers can be determined proceeding from the initial filling degree φ_B.

In case of a hexagonal packing of the fibers in the molded article, to minimize the gap between the fibers, the filling degree is $\varphi_1 = 0.907$.

The geometrical characteristics relating the interfiber distance e to filling degree is $-\varphi_i = 1/(1+e/d_f)^2$. Hence, the initial filling degree can be found so

$$\varphi_B = \varphi_1 \cdot \varphi_i \tag{5.12}$$

By taking the relations above into account, the distance between neighboring fibers can be calculated

$$e = d_f(\sqrt{\varphi_1/\varphi_B} - 1) \tag{5.13}$$

However, this equation does not fully correspond to the filtration conditions of the binder because no allowance has been made for variations in the channel size as binder is removed. This drawback can be partially eliminated by substituting φ_B for $\varphi_{B'} = (\varphi_B + \varphi_E)/2$ in Eq. (5.13). As a result, we then obtain

$$e = d_f\left[\sqrt{\frac{2\varphi_1}{\varphi_B' + \varphi_E}} - 1\right] \tag{5.14}$$

The number of channels can be determined from the equations[55,201]

$$n_r = \frac{\pi \cdot d_0 t}{(d_f + e)^2} = \frac{2\sqrt{3} \cdot d_0 t \cdot \varphi_B}{d_f^2} \tag{5.15}$$

By making use of Eq. (5.6) and upon determining the area of the channels, we can find the amount of the binder that has moved in a radial direction in time corresponding to the winding pitch

$$Q = \frac{4\sqrt{3} \cdot K_p^r t^2 T_0 \varphi_B d_0}{V_S \cdot \eta \cdot L_0'}\left(\sqrt{\frac{2\varphi_1}{\varphi_B + \varphi_E}} - 1\right)^2 \tag{5.16}$$

Here, L_0' is the length of the channels equal to the resultant material thickness. To calculate the permeability factor in the radial direction K_p^r, we presume that $L = d_f/2$ and that the degree of tortuosity of the reinforcement can be found as

$$\frac{L}{L_Q} = \frac{(d_f + e)3\sqrt{3}}{2\pi(d_f + e)} = \frac{3\sqrt{3}}{2\pi} \tag{5.17}$$

Hence,

$$K_{\mathrm{p}}^{\mathrm{r}} = \frac{27(1-\varphi_{\mathrm{B}})^2 d_{\mathrm{f}}^2}{32K_6\varphi_{\mathrm{B}}^2\pi^2}$$

If we substitute Q_{L}, Q_{r}, and Q in Eq. (5.8), the pressure emerging in the winding structure can be determined

$$T_0 = \frac{18\eta K_6 V_{\mathrm{p}}\pi^2 L_0 L_0' \varphi_{\mathrm{B}}^2 (d_0^2 - d^2)[2\varphi_{\mathrm{E}} - \varphi_{\mathrm{B}}(1+\varphi_{\mathrm{E}})]}{\pi^2 L_0'(d_0^2 - d^2)(1-\varphi_{\mathrm{B}}) + 243\sqrt{3}L_0 t\varphi_{\mathrm{B}} d_0 \left(\sqrt{\frac{2\varphi_1}{\varphi_{\mathrm{B}}+\varphi_{\mathrm{E}}}} - 1\right)^2} \tag{5.18}$$

By using the pressure value in the winding structure found with Eq. (5.18), we can then calculate the tension force of the winding reinforcement needed to ensure the required filling degree and dimensions of the article at the winding stage

$$N_0 = \frac{1}{2} T d_0 t \sqrt{1 + \tan^2\gamma} \tag{5.19}$$

The following occasions may arise at the considered stage:

1. The pulling velocity surpasses that of the binder in reverse longitudinal direction. In this case, because of the pressure arising in the structure, the binder will be squeezed through the side surface.

The pressure value can be determined from the parity of the amount of binder removed from the structure and having passed only in the radial direction, as determined by Eqs. (5.9) and (5.16)

$$T' = \frac{2\pi^3}{27\sqrt{3}} \frac{V_{\mathrm{p}}\eta L_0' K_6 \varphi_{\mathrm{B}}(d_0^2 - d^2)(\varphi_{\mathrm{E}} - \varphi_{\mathrm{B}})}{t d_{\mathrm{f}} d_{\mathrm{f}}^2 \varphi_{\mathrm{E}}(1-\varphi_{\mathrm{B}})\left(\sqrt{\frac{2\varphi_1}{\varphi_{\mathrm{B}}-\varphi_{\mathrm{E}}}} - 1\right)^2} \tag{5.20}$$

Based on the found displacement velocity of the binder in the radial direction and taking account of the area of the channels, we obtain an equation for the flow of the binder through the channels formed by the longitudinal and transversal reinforcements

$$q_{\mathrm{r}} = \vec{V}_{\mathrm{r}} F_{\mathrm{k}}' = \frac{2\sqrt{3}K_{\mathrm{p}}^2 T' d_0 t \varphi_{\mathrm{B}}}{\eta L_0'} \left(\sqrt{\frac{2\varphi_1}{\varphi_{\mathrm{B}} - \varphi_{\mathrm{E}}}} - 1\right) \tag{5.21}$$

Let us now determine the time during which this excess binder will move outward in the surface.

$$\tau = \frac{Q}{q_{\mathrm{r}}} \tag{5.22}$$

Once we have determined the pulling velocity, we can find the length of the zone over which pressure of the binder lowers down to atmospheric pressure

$$L = V_S\tau = \frac{\pi}{8\sqrt{3}} \frac{(d_0^2 - d^2)\,\eta\, L_0'(\varphi_E - \varphi_B)\, V_S}{K_{pp}^r\, T' d_0\, \varphi_B\, \varphi_E \left(\sqrt{\frac{2\varphi_1}{\varphi_B - \varphi_{Ek}}} - 1\right)^2} \tag{5.23}$$

In presumption of linear changes in pressure over length e, we can determine the normal load transmitted onto the inner mandrel

$$H = T_0' \frac{\pi d L}{2} \tag{5.24}$$

Now that we have the cohesion factor of the material and the mandrel, we can find the resistance force to motion over this zone

$$P_2 = H f_{fr} + p_0 A \tag{5.25}$$

Relation (5.25) represents the friction law, where p_0 is the adhesion factor, A is the actual contact area of the surfaces, and f_{fr} is the friction coefficient.

The adhesion factor is conditioned by the shifting stresses that arise at relative displacements of contacting surfaces. As the material moves over the mandrel, stresses appearing due to the binder viscosity can be determined by Newton's law as the specific resistance force to the viscous flow

$$p_0 = \tau_{at} = \eta \frac{dV}{dx} \tag{5.26}$$

where η is the dynamic viscosity factor and dV/dx is the velocity gradient.

At constant velocity, the gradient can also be taken as constant, then

$$p_0 = \tau_{at} = C_1 \eta \tag{5.27}$$

In this relation, the dynamic viscosity η is a function of temperature, while factor C_1 is dependent on pulling velocity.

The dynamic viscosity is calculated[26] as follows:

$$\eta = A(T_k) \exp \frac{W}{K \cdot T_k} \tag{5.28}$$

where $A(T_k)$ = some temperature function
W = activation energy of the viscous liquid flow
K = Boltzmann's constant

Following Ref. 26, the activation energy of the viscous flow is

$$W = B_{\mathrm{q}} + \gamma_{\mathrm{q}} - \alpha(T_{\mathrm{k}} - T_{\mathrm{k1}}) \tag{5.29}$$

where B_{q}, γ_{q}, and α = constants
q = liquid pressure
$(T_{\mathrm{k}} - T_{\mathrm{k1}})$ = temperature variation

As it follows from Eqs. (5.27) and (5.28), the viscosity will decrease with increasing liquid temperature but will increase with growth in pressure, q.

In stationary conditions, the outer friction force is constant;[201] consequently

$$\eta = A(T_{\mathrm{k}}) \exp \frac{B_q + \gamma_q \cdot q}{K \cdot T_{\mathrm{k}}} \tag{5.30}$$

Hence, from Eqs. (5.27) and (5.30) we can find p_0

$$p_0 = A(T_{\mathrm{k}}) C_1 \exp \frac{B_{\mathrm{q}} + \gamma_{\mathrm{q}} \cdot \mathrm{q}}{K \cdot T_{\mathrm{k}}} \tag{5.31}$$

Relation (5.31) helps to qualitatively define the effect of the principal parameters on the adhesion factor. However, finding the constituent values of Eq. (5.31) in practice is rather difficult. In this connection, experimental investigations have been undertaken to determine the adhesion factor taking into account the material motion velocity and temperature. A scheme of an experimental setup is illustrated in Fig. 5.2.

A preliminary prepreg, prepared and filled to a needed degree, has been drawn using the active grip of a testing machine through a

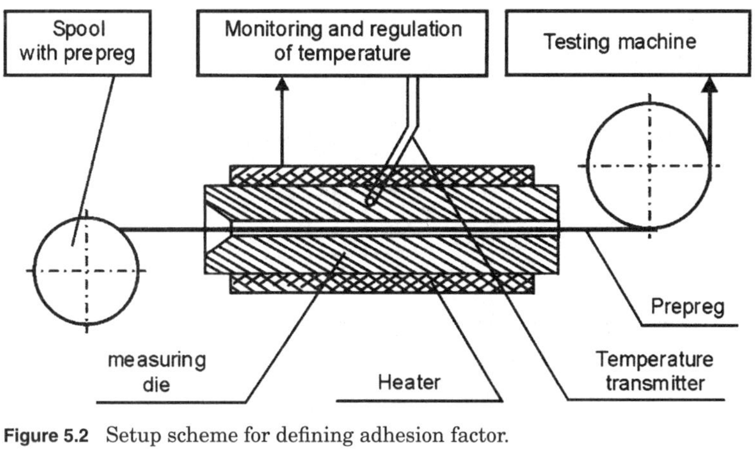

Figure 5.2 Setup scheme for defining adhesion factor.

measuring die of a given diameter and measurable length. Prior to measurements, to avoid the reduction in pressure and ensure the needed binder content and size, the prepreg was drawn through a calibrating die. As the resistance to motion through the die is generally a sum of the resistance of the die inlet, resistance to sliding friction, and the resistance brought about by adhesion, all measurements were then performed in dies of different lengths while the first two forces were excluded, as all other parameters were equal.

Bearing in mind that, in general, the resistance force is defined by Eq. (5.25), one can find the adhesion factor by measuring the resistance forces for two die area values

$$p_0 = \frac{F_2 - F_1}{A_2 - A_1} \tag{5.32}$$

where F_2 and F_1 are pulling force values of the prepreg on die areas A_2 and A_1, respectively.

It is natural that such parameters as area, temperature, and motion velocity affect factor p_0 in a certain manner. The experimental results on the effect of pulling velocity at different temperatures of dies are presented in Fig. 5.3.

It is evident that, as the velocity increases, the adhesion factor decreases for all die temperature values. Changes in the die area leads but only initially to feeble p_0 growth (Fig. 5.4), which in fact then remains invariable for a given range of die temperatures.

To estimate the effect of the prepreg filling degree on the pulling force, different amounts of reinforcing fillers were introduced into the die at constant open flow area, for which the filling degree was preliminarily calculated. The experimental results are presented in Fig. 5.5.

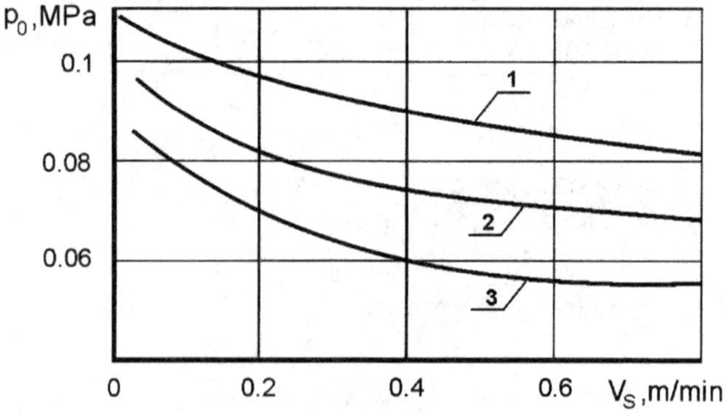

Figure 5.3 Adhesion factor versus motion velocity of prepregs at different die temperatures: (1) $T = 20°C$, (2) $T = 40°C$, (3) $T = 60°C$.

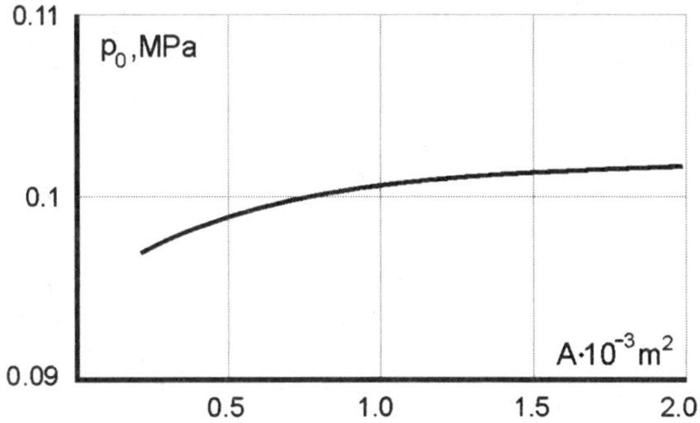

Figure 5.4 Adhesion factor versus die area.

The data presented shows that the pulling force is augmented over the range $0.35 \leq \varphi \leq 0.7$ (Fig. 5.5). With further increment in the filling degree, the resistance force value is observed to drop because of the smaller actual adhesion area and intensified competing process of dry friction. The approximation curve can be described by a regression relationship

$$F/F_0 = -0.0237\varphi^3 + 0.353\varphi^2 - 1.53\varphi + 3.062 \qquad (5.33)$$

where F_0 is the force value at filling degree of the material $\varphi = 0.3$.

2. The pulling velocity is lower with the reverse flow rate of the binder. In this case, the binder will be fully squeezed into the region in front of the winding zone. The pressure can be defined here based on the magnitude of the binder flow in the longitudinal direction

$$q_L = (V_L - V_S)F_k'' = \frac{\pi\left(d_0^2 - d^2\right)}{4}\left(\frac{T'' K_p^L}{\eta L_0} - V_S\right)(1 - \varphi_B) \qquad (5.34)$$

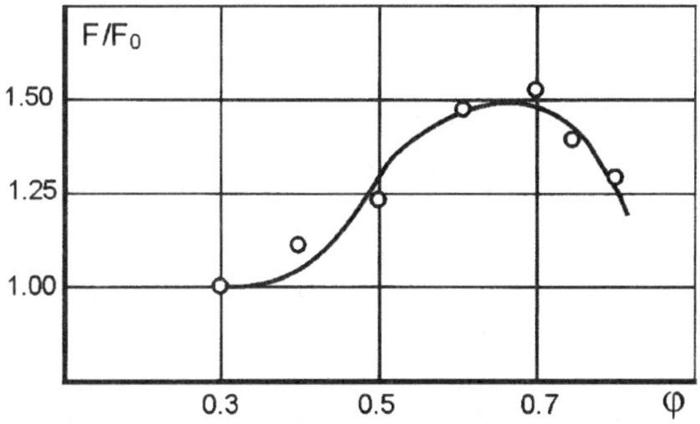

Figure 5.5 Relative alteration of the prepreg pulling force through the measuring die depending on reinforcing degree.

Let us find the value of pressure T'' proceeding from the equality of the removed binder volume and that of the flowing in longitudinal direction by Eqs. (5.9) and (5.11)

$$T'' = \frac{16\eta\, L_0 K_6\, V_S \varphi_B^2 [2\varphi_E - \varphi_B(1-\varphi_E)]}{d_f^2\, \varphi_E(1-\varphi_B)} \tag{5.35}$$

This pressure is generated over a length equal to winding pitch t, which is then reduced over some length to zero, where the smallest length value is also equal to the winding pitch. Such variations of the pressure can be estimated in principle in terms of the parity of the binder volume removed (Eq. (5.9)) and the binder issuing in the longitudinal and transversal directions under the initial pressure effect. We can determine the pressure using Eqs. (5.11) and (5.16), in which $L_0 = t$ and permeability factors K_p^T and K_p^L make allowances for the material-free area outside the winding zone and for the channel sizes in the radial direction, $t{\cdot}e$. The calculations show that the pressure variation T'' over the described area also behaves linearly. Hence, to estimate the resistance force to motion, one can employ relation (5.36), which is analogous to Eq. (5.25)

$$P_2' = \pi T'' t \cdot d \cdot f_{fr} + p_0 A \tag{5.36}$$

Apparently, it is feasible to create a needed material by the technological process under study in case the binder will be displaced in the course of being squeezing in the radial direction only. This circumstance is required because, for the cross-underwinding, one may employ preimpregnated prepregs, whereas for a layer of braided reinforcement only dry braids are used for impregnation, for which a definite amount of the binder must be found on the side surface. The critical value of the pulling force, lower than which it becomes impossible to impregnate the whole sleeve, can be determined as follows

$$V_{\lim} \geq \frac{(1-\varphi_B)^2 d_f^2 N_0}{8\eta\, K_6\, t^2 \varphi_B\, d_0 \sqrt{1-\tan^2\gamma}} \tag{5.37}$$

Analysis of the dependence presented above has shown that the principal structural, geometrical, force, and technological parameters are highly sensitive to variations and should be treated with carefully. Note that aiming at increasing the limiting velocity and, thereby, on the face of it, raising process efficiency may result in a significant growth of the resistance force (Eqs. (5.20) and (5.25)).

5.3 Determination of Resistance Force to Pulling in the Braiding Zone

It is to be emphasized that in contrast to cross winding the angles between braids in the braided zone are larger and braiding is made by a greater number of carriers. Formed by the braiding filaments surface has been described in Chap. 2.

To consider operating loads exerted by a single carrier, the calculation model can be presented as illustrated in Fig. 5.6.

Bear in mind that the angle between the tension stress projection of the braiding reinforcement and the y–γ axis is rather large at braiding; then, firstly, the force $P_{4'}$ governed by the tension stress projection on the article long axis and, secondly, the molding pressure of the article T_p will be exerted from the side of the braiding threads due to the radial force constituent like at cross-winding.

The mentioned values can be calculated as follows:

$$(P'_4)_1 = P'_\mathrm{B} \cos\alpha_1 \sin\gamma \tag{5.38}$$

$$T_\mathrm{p} = \frac{n \cdot P'_\mathrm{B}}{d_0 t_\mathrm{b} \sqrt{1 + \tan^2\gamma}} \tag{5.39}$$

where n = number of braiding reinforcement carriers
t_b = braiding pitch
d_0 = outer diameter of the inner layer of the article

Remembering that half of the carriers are found on the outer diameter of the braider and the second half on the inner diameter, thus

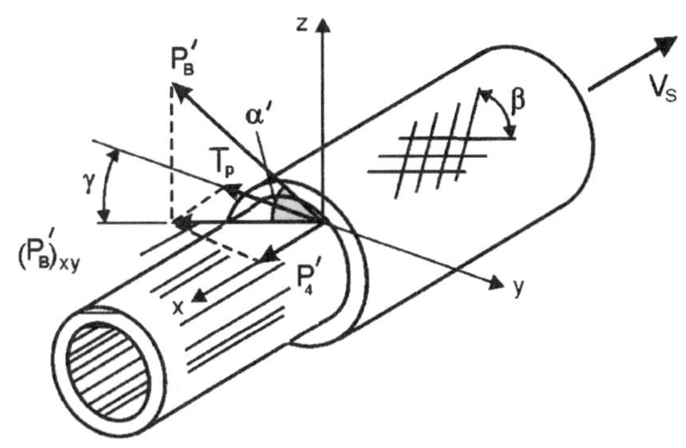

Figure 5.6 Scheme of operating forces at braiding: x, y, z, local coordinates; $(P_{B'})_{XY}$, projection of $P_{B'}$ force on plane xoy.

diminishing the tension stress, and assuming a linear alteration in tension, we can obtain from Eq. (5.38)

$$P_4' = \frac{n}{2}(P_4')_1 \tag{5.40}$$

Simultaneously, taking into account that the material's inner structure has been shaped at the previous stage of braiding and cannot be deformed, the binder squeezed from the structure during cross-winding then filters through the braided layer. The amount of the binder can be estimated within the braiding pitch as

$$Q = \frac{\pi}{4}(d_b^2 - d_o^2)t_b(1 - \varphi_I)k_Z \tag{5.41}$$

where d_b = article diameter after braiding
φ_1 = filling degree of the ready article
k_Z = accepted load factor (k_Z = 1.2–1.3)

As discussed earlier, the amount of the binder able to permeate through the braided layer is found as follows:

$$Q_r = \vec{V} S_C \tau \tag{5.42}$$

where $\vec{V}$ = mean velocity of the liquid phase
S_C = area of channels in the radial direction
τ = time during which the binder is squeezed

Presuming that the binder is squeezed only in the radial direction as it is the most expedient one, the distance between threads can be found from Eq. (5.14), and the number of cannels from

$$n_r = \frac{\pi \cdot d_0 t_b}{(d_f + e)^2 \sin\beta} \tag{5.43}$$

where β is the angle between threads in a semi (Fig. 5.6).

Using Eqs. (5.14) and (5.43), we can find S_C and the amount of the binder squeezed in radial direction

$$Q_r = \frac{K_p^r T}{\eta L_r} \frac{\pi \cdot d_0 t_b e^2}{(d_f + e)^2} \tau \tag{5.44}$$

To calculate the permeability factor in the radial direction, K_p^r, we presume that $L/L_q = 1$, $\varphi_n = \varphi_I(d_2^2 - d_0^2)/(d_b^2 - d_0^2)$, $U = 1 - \varphi_n$, $S_v = 4\varphi_n/d_f$, and $L_r = (d_b - d_0)/2$; d_2 is the diameter of the final article.

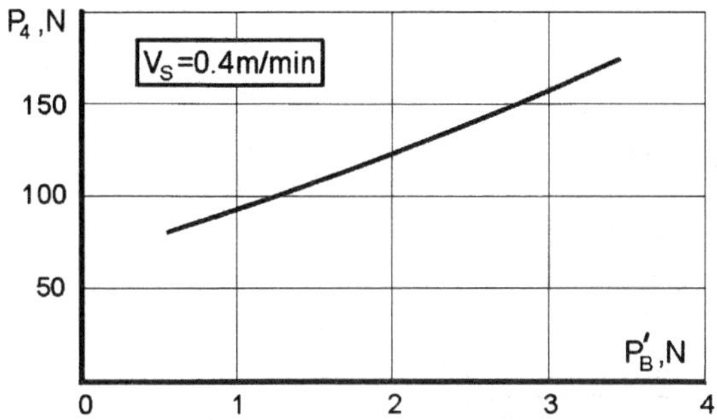

Figure 5.7 Variations in P_4 force depending on tension force of the braiding reinforcement.

By equating Eqs. (5.44) and (5.41), the time during which the binder penetrates through the braided layer can estimated

$$\tau = \frac{K\eta\,(d_b^2 - d_0^2)(d_b - d_0)(d_f + e)^2}{8K_p^r T d_0 e^2} \tag{5.45}$$

Based on known τ, one can define the semi-length over which pressure T is effective

$$L' = V_S \tau = \frac{V_S K\eta\,(d_b^2 - d_0^2)(d_b - d_0)(d_f + e)^2}{8K_p^r T d_0 e^2} \tag{5.46}$$

Anticipating the linear pressure distribution, analogously to Eq. (5.25), we can find the resistance force to motion in zone IV

$$P_4'' = \frac{TL'\pi \cdot d}{2} f_{fr} + p_0 \cdot \pi \cdot d \cdot L' \tag{5.47}$$

Finally, we can calculate the total resistance force in zone IV

$$P_4 = P_4' + P_4'' \tag{5.48}$$

A part of calculation results of force P_4 is illustrated in Figs. 5.7 and 5.8. As the analysis proves, increasing tension stress of the braiding reinforcement leads to an augmented resistance force. Similar variations take place with increasing reinforcement angle. The greatest effect on the force P_4 is imposed by the adhesive constituent of the semi-finished product onto the mandrel surface.

5.4 Investigations of Die Entry Block Parameters

Along with the analysis of the binder pressure generated in the die,[131,133,297,298] the effects in its entry block and molding conditions

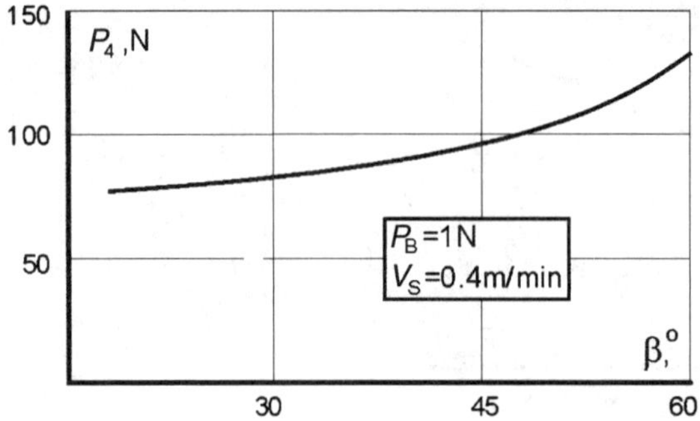

Figure 5.8 Variation of P_4 depending on reinforcement angle.

have been comprehensively studied.[119,131,296,317–319] It has been established that inlet pressure and die parameters are of paramount importance for the quality of products obtained.

It is possible to compose a balance of displaced volumes of the material in terms of an elementary volume of the material molded (Fig. 5.9).

$$\tau\,[q_{\mathrm{L}}(x_1) - q_{\mathrm{L}}(x)] = dW \tag{5.49}$$

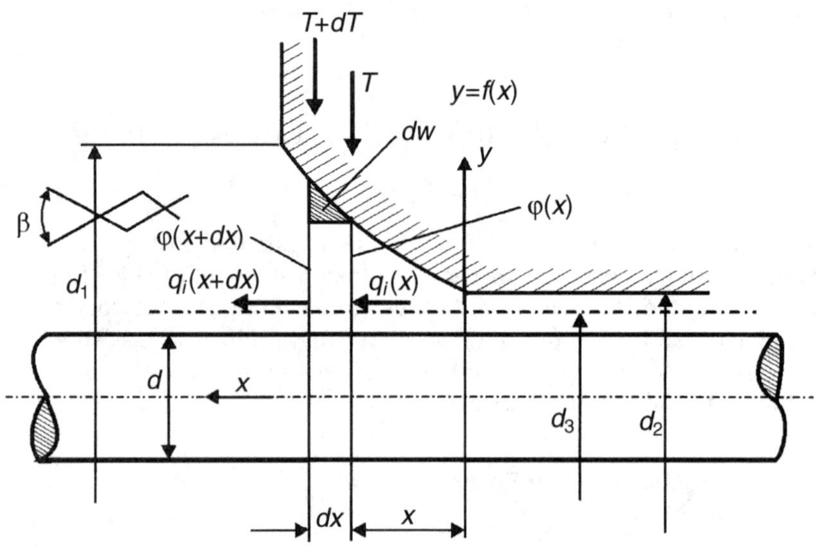

Figure 5.9 Squeezing of the binder within the die entrance.

where $q_L(x_1)$ = flow of the binder through the section at a distance $x_1 = x + dx$ from the point of origin of the die entry block with a forming channel

$q_L(x)$ = flow through the section at a distance x

τ = time of article motion

dW = volume of the binder filtered out during motion

Flows of the binder squeezed through the cross-sections are calculated by the following relations:

$$\begin{aligned} q_L(x_1) &= \vec{V}'_L S'_C \\ q_L(x) &= \vec{V}''_L S''_C \end{aligned} \tag{5.50}$$

As previously, flow rates of the binder are calculated by relation (5.6), where the permeability factor is found from Eq. (5.7). During motion of the binder in the die in which the temperature alters lengthwise, the viscosity becomes a value that is dependent upon location in the section. In accordance with the assumptions accepted earlier, the parameters included in the velocity relation are estimated as follows:

$$\begin{aligned} V_i &= 1 - \varphi(x_i) \\ S_{vi} &= 4\varphi(x_i)/d_f \\ L_i &= dx/\sin\beta/2 \end{aligned} \tag{5.51}$$

The open flow area values included in Eq. (5.50) are calculated by the formula

$$S_{ki} = \left[y^2(x_i) - r_3^2\right]\left[1 - \varphi(x_i)\right] \tag{5.52}$$

and the filling degree of the current section will be

$$\varphi(x_i) = \varphi(x)\frac{y^2(x) - r_3^2}{y^2(x_i) - r_3^2} \tag{5.53}$$

We denote parameters of further sections with i, while the angle β is the angle of reinforcement of the outer layer.

It is necessary to set the shape of the entry surface $y = f(x)$ in all relations used. For example, if the die entry block shape is considered to be parabolic, as is most rational, then

$$y = ax^2 + b \tag{5.54}$$

we can obtain a relation for volume dW in the form

$$dW = \pi \left[\int_{X}^{X_1} y^2(x)dx - y^2(x)dx \right] \tag{5.55}$$

Hence, by substituting found formulas into Eq. (5.51) and anticipating that $\tau = dx/V_S$, we can derive the next relation for the pressure in volume dV in accordance with the accepted law of variation in channel shape

$$T(x_i) = \frac{\eta V_S L_i}{dx S_{k2} K_p''} \left\{ \frac{dx V_L' S_{k1}}{V_S} + \frac{a^2 x_i^2}{5} + \frac{2abx_i}{3} \right.$$
$$\left. + b^2 x_i - \frac{a^2 x^5}{5} - \frac{2abx^3}{3} - b^2 x - (ax^2 + b)^2 dx \right\} \tag{5.56}$$

In the latter relation, $x = 0$, which means that at the end of the entry block the filling degree in the channel is $\varphi(x) = \varphi_E$, i.e., it is equal to the filling degree of the article, while the pressure value, $T(x = 0)$, is conditioned by the material volume expansion as a result of heating to the gelation temperature. It is also possible to assert that the viscosity of the binder during its displacement along the entry part of the channel changes linearly

$$\eta(x) = kx + \eta_0 \tag{5.57}$$

where η_0 is the viscosity in section $x = 0$ and the coefficient $k = 0.2 \cdot 10^3$ Pa·s/m.

The constituent of the resistance force to motion that is dependent upon the die reaction is found from the relation

$$P_5' = \int_{0}^{X_k} T(x_i) \Delta S \sin \theta_i dx \tag{5.58}$$

Angle θ_i is calculated here in each point of the die by obeying the accepted law of its shape variation in a point with coordinate $z = x_i = dx/2$.

Hence, from Eq. (5.56) we obtain

$$\theta_i = \arctan(2az) \tag{5.59}$$

The area ΔS on the die portions of length dx is expressed by the formula

$$\Delta S = 2\pi \int_{X}^{X_1} y(x)\sqrt{1 + (y'(x))^2} dx \tag{5.60}$$

Under the accepted variation law (Eq. (5.56))

$$\Delta S = 2\pi \left[\frac{1}{8a^2}(A_1 - A_2) + \frac{b}{2a}(C_1 - C_2) \right] \tag{5.61}$$

where

$$A_k = \frac{ax_n}{2}\sqrt{[1+(2ax_n)^2]^3} - \frac{1}{8}\left[2ax_n\sqrt{1+(2ax_n)^2} + \ln(2ax_n + \sqrt{1+(2ax_n)^2}\right]$$

$$C_k = \frac{1}{2}\left[2ax_n\sqrt{1+(2ax_n)^2} + \ln(2ax_n + \sqrt{1+(2ax_n)^2}\right]$$

An additional constituent of the resistance force arising from the material friction against the die and mandrel walls is calculated by the equation

$$P_5'' = f_{fr}\left[\int_0^{X_k} T(x_i)\,\Delta S\cos\theta_i\,dx + \frac{\pi d}{2}\int_0^{X_k} T(x_i)\,dx\right] + P\left[\int_0^{X_k} \Delta S\cos\theta_i\,dx + \frac{\pi d}{2}x_k\right] \tag{5.62}$$

Coordinate x_k defines here the section corresponding to the contact moment of the article outer surface with the die.

Taking account of the variation law (Eq. (5.60)), we have

$$x_k = \sqrt{\frac{r_1 - b}{a}} \tag{5.63}$$

Hence, the resultant resistance force to motion in the die zone will be

$$P_5 = P_5' + P_5'' \tag{5.64}$$

For numerical evaluation of the results obtained, the iteration method has been employed, according to which the iteration step t_a has been chosen and the meanings of all relations calculated under argument values x and $x_i = x + t_a$.

In Fig. 5.10, a diagram of the pressure distribution along the length of the die is presented in which the forming channel obeys the function $y = 60x^2 + 11.5 \cdot 10^{-5}$. The analysis of this relation has proved that the pressure distribution is non-uniform along the length of the entry block and that it reaches a maximum at a distance $0.65x_k$. Variations in the resistance force under the action of the die reaction are illustrated in Fig. 5.11.

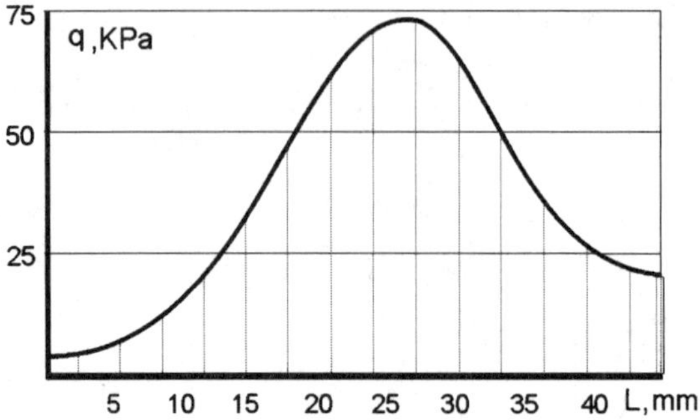

Figure 5.10 Pressure distribution diagram along the entry block length.

Specific interest represents evaluation of the forming channel shape effect on alterations of the process force parameters. In Fig. 5.12 is shown a diagram of the maximum pressure variation q in the material and the total resistance force to motion P_5 in response to changes in the "a" parameter within the die shape variation law (Eq. (5.56)).

An increase in the a value is seen to diminish contact length in the die and to ambiguously affect the parameters under study. In contrast to maximum pressure, which increases monotonously with an increase in a, the total force P_5 displays a minimal value around $a = 55$. This is because the contact length in the forming zone contracts as a increases, whereas pressure intensifies and the constituent of force P_5' is augmented. As a diminishes, the pressure varies only negligibly, while the constituent P_5'' grows essentially as a result of the increased contact zone. Proceeding from the competing processes affecting the total resistance force to motion, it is possible to select the shape of the entry block with account of the contact zone and die surface.

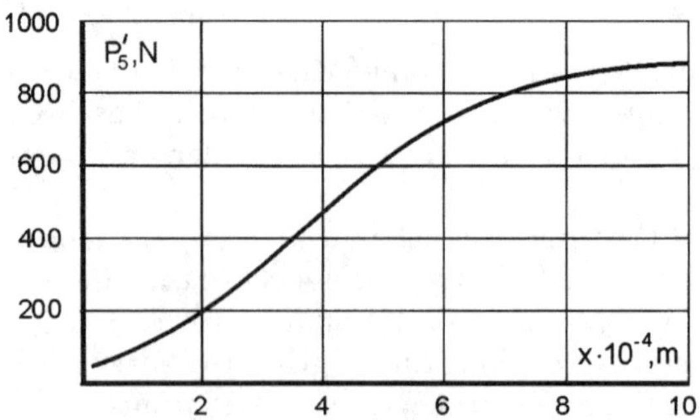

Figure 5.11 Variation of the resistance force constituent to motion P_5' lengthwise along the entry block.

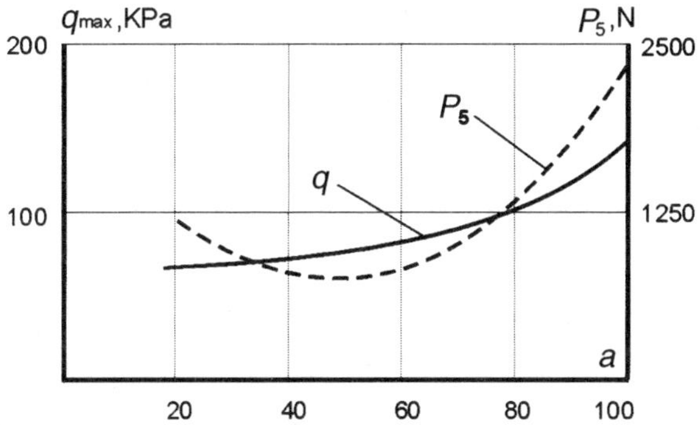

Figure 5.12 Variation in maximum pressure in the entry block q and resistance force P_5 depending on a.

Analysis of above dependencies shows that, when a group of dies is used, each of which is intended to partially mold the article, the resistance force can be lowered. Relevant calculations have shown that for an optimum profile of the die at pulling an article having the initial filling degree of the braided layer $\varphi_B = 0.24$, throughput rate 0.4 m/min, and inner diameter of the article 19.5 mm, a total force $P_5' = 1171.5$ N is generated. When two dies are used with corresponding conformity of diameters, it is possible to lower the force down to $P_5' = 283 + 664 = 947$ N. In this case, the first forming die will show a lower resistance force value due to its lower filling degree.

The dependencies of resistance forces to motion upon major process parameters are presented in Figs. 5.13 to 5.15, i.e., the cross-sectional value, throughput rate, and final filling degree. Based on the information cited, the technological process regimes can be adjusted.

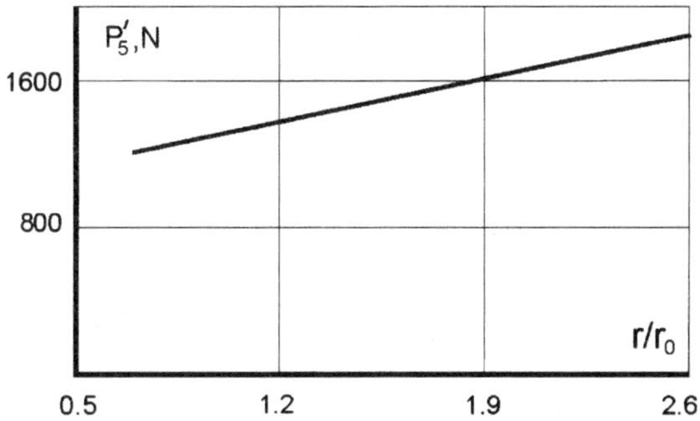

Figure 5.13 Force P_5' dependence on relative cross-section.

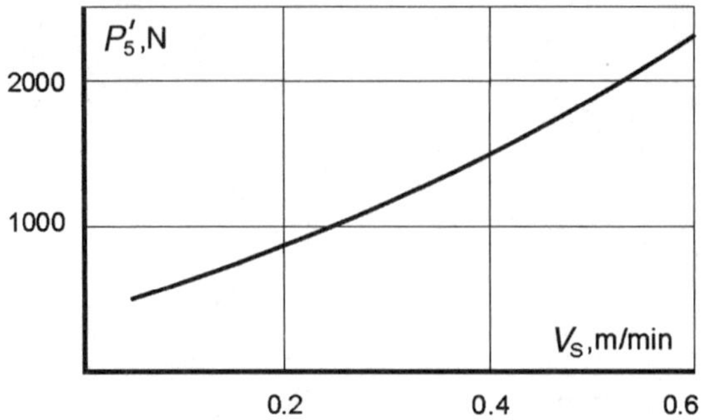

Figure 5.14 Force P_5' alteration depending on throughput rate.

The backflow of the binder squeezed from the entry under the overpressure effect (Fig. 5.16) generates an additional resistance force. To estimate its value, one should define the pressure variation law along the article length at leaving the die. In this case, we consider that the binder is displaced by pressure at the entrance in only the radial direction (Eq. (5.56)).

The flow of binder passing through the section at distance x can then be found as

$$q_{\mathrm{L}}(x) = \frac{K_{\mathrm{p}}^{\mathrm{L}}\, T(x)}{\eta x} S_{\mathrm{C}} \tag{5.65}$$

and at distance $(x + dx)$ as

$$q_{\mathrm{L}}(x + dx) = \frac{K_{\mathrm{p}}^{\mathrm{L}}\, T(x)\, S_{\mathrm{C}}}{\eta x} + \frac{K_{\mathrm{p}}^{\mathrm{L}} dT\, S_{\mathrm{C}}}{\eta x} \tag{5.66}$$

Note that a part of the binder is displaced in radial direction

$$q_{\mathrm{r}}(x) = \frac{2K_{\mathrm{p}}^{\mathrm{r}}[T + dT/2]\, S_{\mathrm{C}}'}{\eta d} \tag{5.67}$$

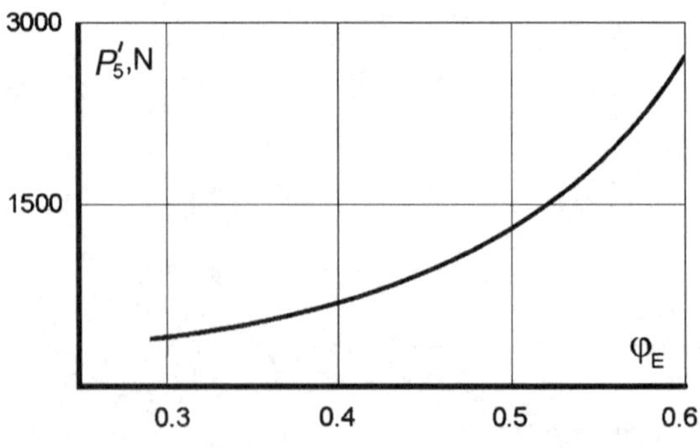

Figure 5.15 Force P_5' dependence on final filling degree of the article.

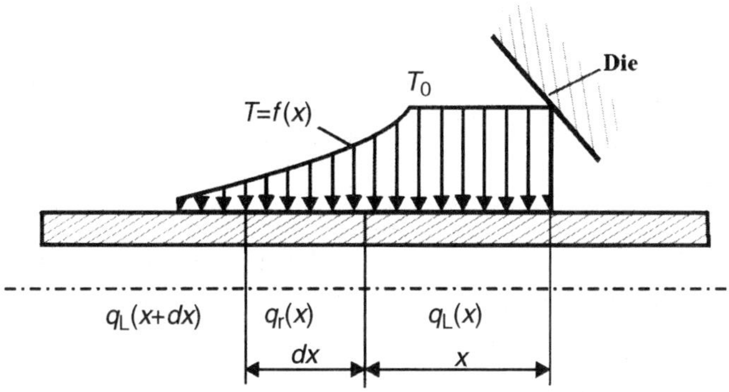

Figure 5.16 Pressure gradient variation at die outlet.

Presuming that the area of channels S_C' is linearly dependent on the size of straight-through channels $S_C' = n_r e_1^2$ in zone dx, we can find the number of channels

$$n_r = \frac{2\sqrt{3}}{d_f^2} d\varphi_B \, dx \tag{5.68}$$

By substituting $e_1 = d_f(\sqrt{\varphi_1/\varphi_B} - 1)$ and n_r from Eq. (5.68) and remembering that $K_p^L/K_p^r = 27(4\pi^2) = 0.68$, we obtain

$$\frac{dT}{T(x)} = -\frac{6\left(\sqrt{0.907/\varphi_B} - 1\right)^2 x dx}{d^2(1-\varphi_B)} \tag{5.69}$$

The solution of this equation is

$$T(x) = C \exp(-m_1 x^2) \tag{5.70}$$

where $m_1 = \frac{3(\sqrt{0.907/\varphi_B}-1)^2}{d^2(1-\varphi_B)}$ The constant C is determined with regard to original conditions; at $x = 0$ (Fig. 5.16), $C = T_0$.

We can now find the volume of the binder removed from the article,

$$\tau = \frac{L_x}{V_S} \tag{5.71}$$

in time equal to the total volume of the binder passing through the article surface

$$Q_r = \tau \int_0 \frac{4\sqrt{3}K_p^r \varphi_B}{\eta} \left(\sqrt{0.907/\varphi_B} - 1\right)^2 T(x) dx \tag{5.72}$$

Bearing in mind that $\int_0^\infty e^{-m_1}x^2 = \frac{\sqrt{\pi}}{2\sqrt{m_1}}$, we then obtain

$$Q_r = \frac{2\sqrt{\pi}K_p^r dL_x T_0 \varphi_B \sqrt{1-\varphi_B}}{V_S \eta}\left(\sqrt{0.907/\varphi_B} - 1\right)^2 \tag{5.73}$$

To determine the length L_x over which the pressure $T(x)$ operates, we can equate the binder volume found by Eq. (5.73) and the volume of the binder removed from the die determined earlier by Eq. (5.55)

$$L_x = \frac{V_S \eta \sqrt{\pi}\left[\int_0^{X_k} y^2(x)dx - y^2(x)dx\right]}{2K_p^r dT_0 \varphi_B \sqrt{1-\varphi_B}(\sqrt{0.907/\varphi_B} - 1)^2} \tag{5.74}$$

Then, with Eq. (5.72), we may find the resistance force to pressure

$$P_5''' = \pi dL_x \int_0^{L_x} T(x)dx f_{fr} \tag{5.75}$$

This force should be naturally included in relations (5.66) for determining P_5.

5.5 Investigations of Resistance Force During Motion in the Die

An article is known to be solidified as the molded material moves within the die. This process goes hand-in-hand with conversion of the binder from the liquid state to the solid state. This is accompanied by its shrinkage and changes in dimensions at thermal expansion. To estimate the resistance force to motion, it is necessary to determine the pressure magnitude exerted on the die. A series of works have been devoted to pressure studies in the die during molding of various products,[118,119,131,296–298,318] in which the pressure value and its effect on the physicomechanical characteristics of products obtained have been examined. Because of the varying aggregate state of the material along the length of the die during solidification, its friction coefficient and pressure vary, too, but the limits and extension of these variations are very hard to define. In this connection, to simplify the calculations the named parameters are taken to be invariable. To find the pressure values, let us consider a glass-reinforced plastic shell with a required structure that, on the one hand, changes its dimensions as a result of shrinkage at solidification and, on the other, expands under thermal effects.[116]

Strains that emerge during these processes can be compensated for by application of certain inner pressures of corresponding magnitude.

Presuming that the considered material shows in tangential direction a linear thermal expansion factor α_θ and shrinkage ξ, we can determine relative deformation in the tangential direction

$$\varepsilon_\theta = \xi(1-\varphi) - \alpha_\theta \Delta T \tag{5.76}$$

Let us now take an anisotropic shell experiencing the effect of some inner pressure. In the general case, when solving the deformation problem of a plane shell with a cylindrical anisotropy, we can write the equilibrium equation as

$$\frac{\partial \sigma_r}{\partial r} + \frac{\sigma_r - \sigma_\theta}{r} = 0 \tag{5.77}$$

Thus, the equations for the strain to stress relation will be

$$\begin{aligned}
\varepsilon_r &= \frac{\sigma_r}{E_r} - \frac{\nu_{\theta r}}{E_\theta}\sigma_\theta - \frac{\nu_{zr}}{E_z}\sigma_z \\
\varepsilon_\theta &= \frac{\nu_{r\theta}}{E_r}\sigma_r + \frac{\sigma_\theta}{E_\theta} - \frac{\nu_{z\theta}}{E_z}\sigma_z \\
\varepsilon_z &= \frac{\nu_{rz}}{E_r}\sigma_r - \frac{\nu_{\theta r}}{E_\theta}\sigma_\theta + \frac{\sigma_z}{E_z}
\end{aligned} \tag{5.78}$$

The relation of the strain to stress constraints is of the form

$$\varepsilon_r = \frac{\partial U_{\mathrm{r}}}{\partial \mathrm{r}}; \qquad \varepsilon_\theta = \frac{U_r}{r}; \qquad \varepsilon_z = 0 \tag{5.79}$$

By substituting $\varepsilon_z = 0$ in Eq. (5.78), we obtain

$$\sigma_z = E_z\left(\frac{\nu_{rz}}{E_r}\sigma_r + \frac{\nu_{\theta z}}{E_\theta}\sigma_\theta\right) \tag{5.80}$$

Let us substitute the above relation into the first two equations of system (5.78)

$$\begin{aligned}
\varepsilon_r &= \frac{\sigma_r}{E_r}(1 - \nu_{zr}\nu_{rz}) - \frac{\sigma_\theta}{E_\theta}(\nu_{\theta r} + \nu_{zr}\nu_{\theta z}) \\
\varepsilon_\theta &= \frac{\sigma_\theta}{E_\theta}(1 - \nu_{z\theta}\nu_{\theta z}) - \frac{\sigma_r}{E_r}(\nu_{r\theta} + \nu_{z\theta}\nu_{rz})
\end{aligned} \tag{5.81}$$

The differential compatibility equation based on Eq. (5.79) is

$$r\frac{\partial \varepsilon_\theta}{\partial r} + \varepsilon_\theta - \varepsilon_r = 0 \tag{5.82}$$

By denoting

$$\frac{E_\theta(1-\nu_{zr}\nu_{rz})}{E_r(1-\nu_{z\theta}\nu_{\theta z})}=\vartheta^2 \tag{5.83}$$

and

$$\frac{\nu_{r\theta}+\nu_{z\theta}\nu_{rz}}{1-\nu_{zr}\nu_{rz}}=\chi \tag{5.84}$$

and substituting Eq. (5.81) in Eq. (5.82), we obtain the equation

$$r\left(\frac{\partial\sigma_\theta}{\partial r}-\vartheta^2\chi\frac{\partial\sigma_r}{\partial r}\right)+\sigma_\theta-\vartheta\chi\sigma_r-\vartheta^2\sigma_r+\vartheta^2\chi\sigma_\theta=0 \tag{5.85}$$

Let us now introduce the stress function in the form $\varphi = r\cdot\sigma_r$. Proceeding from equilibrium condition (Eq (5.77)), we then obtain

$$\sigma_r=\frac{\phi}{r};\qquad \sigma_\theta=\frac{\partial\phi}{\partial r} \tag{5.86}$$

By substituting Eq. (5.86) in Eq. (5.85),

$$\frac{\partial^2\phi}{\partial r^2}+\frac{1}{r}\frac{\partial\phi}{\partial r}-\vartheta^2\frac{\phi}{r^2}=0 \tag{5.87}$$

The differential Eq. (5.78) obeys the function

$$\phi=C_1r^{\vartheta}+C_2r^{-\vartheta} \tag{5.88}$$

So far, stresses σ_r and σ_Q are found as follows:

$$\begin{aligned}\sigma_{\mathrm{r}}&=C_1r^{\vartheta-1}+C_2r^{(-\vartheta+1)}\\ \sigma_\theta&=C_1\vartheta r^{\vartheta-1}-C_2\vartheta r^{(-\vartheta+1)}\end{aligned} \tag{5.89}$$

Constants C_1 and C_2 are calculated from the boundary conditions $\sigma_{\mathrm{r}} = -q$ at $r = r_{\mathrm{I}}$ and $\sigma_r = 0$ at $r = r_{\mathrm{E}}$

$$\begin{aligned}C_1&=\frac{q}{(r_{\mathrm{E}}/r_{\mathrm{I}})^{2\vartheta}-1}\,\frac{1}{\chi^{\vartheta-1}}\\ C_2&=\frac{(r_{\mathrm{E}}/r_{\mathrm{I}})^{\vartheta-1}q}{(r_{\mathrm{E}}/r_{\mathrm{I}})^{2\mathrm{m}}-1}\cdot r_{\mathrm{E}}^{r+1}\end{aligned} \tag{5.90}$$

where r_{I} is the inner radius and r_{E} is the outer radius.

After substitutions of Eq. (5.90) in Eq. (5.89), the relations for stresses will be

$$
\begin{aligned}
\sigma_{\mathrm{r}} &= \frac{q}{(r_{\mathrm{E}}/r_{\mathrm{I}})^{2\vartheta}-1}\left(\frac{r}{r_{\mathrm{I}}}\right)^{\vartheta-1} - \frac{q(r_{\mathrm{E}}/r_{\mathrm{I}})^{\vartheta-1}}{(r_{\mathrm{E}}/r_{\mathrm{I}})^{2\vartheta}-1}\left(\frac{r_{\mathrm{E}}}{r}\right)^{\vartheta-1} \\
\sigma_{\Theta} &= \frac{q}{(r_{\mathrm{E}}/r_{\mathrm{I}})^{2\vartheta}-1}\left(\frac{r}{r_{\mathrm{I}}}\right)^{\vartheta-1}\vartheta + \frac{q(r_{\mathrm{R}}/r_{\mathrm{I}})^{\vartheta-1}}{(r_{\mathrm{E}}/r_{\mathrm{I}})^{2\vartheta}-1}\cdot\left(\frac{r_{\mathrm{E}}}{r}\right)^{\vartheta-1}\vartheta
\end{aligned}
\tag{5.91}
$$

Dependencies (Eq. (5.91)) show distribution of stresses over the radius. If we substitute the found stress values in the second relation of system (5.81), we can calculate the dependence of a relative strain in the tangential direction

$$
\begin{aligned}
\varepsilon_\theta = \frac{q}{A^{2m}-1}\bigg[&\frac{\vartheta}{E_\theta}(B^{\vartheta-1}+A^{\vartheta-1}D^{\vartheta+1})(1-\nu_{z\theta}\nu_{\theta z}) \\
&- \frac{1}{E_r}(B^{\vartheta-1}-A^{\vartheta-1}D^{\vartheta+1})(\nu_{r\theta}+\nu_{z\theta}\nu_{rz})\bigg]
\end{aligned}
\tag{5.92}
$$

where $A = r_{\mathrm{E}}/r_{\mathrm{I}}$; $B = (r_{\mathrm{E}} - r_{\mathrm{I}})/2r_{\mathrm{I}}$; $D = r_{\mathrm{E}}/(r_{\mathrm{E}} - r_{\mathrm{I}})$.

By equating Eqs. (5.92) and (5.76) for tangential strains, we can find the pressure q of the set article on the inner mandrel in the die

$$
q = \frac{[\xi(1-\varphi) - \alpha_\theta \Delta T]\cdot(A^{2\vartheta}-1)}{\left[\frac{\vartheta}{E_\theta}(B^{\vartheta-1}+A^{\vartheta-1}D^{\vartheta+1})(1-\nu_{z\theta}\nu_{\theta z}) - \frac{1}{E_r}(B^{\vartheta-1}-A^{\vartheta-1}D^{\vartheta+1})(\nu_{r\theta}+\nu_{z\theta}\nu_{rz})\right]}
\tag{5.93}
$$

where ΔT is the temperature variation.

The elastic constants and the linear thermal expansion factor included in this relation can be calculated as follows. The composite under consideration represents a laminated material (Fig. 5.17) consisting of a braided layer with fibers arranged at $\pm\alpha$ angle to the article axis, plus one longitudinal layer and one transversal one.

There exist a number of calculation procedures for estimating deformation properties of laminated composites with intricate compound reinforcement schemes. The resultant analytical dependencies are, as a rule, bulky, and difficulties arise at active valuation of the materials' physicomechanical characteristics. A structural model of a composite has been proposed[325] based on which simplified relations for elastic characteristics of the material studied have been derived. The calculation procedure as applied to the material structure is as follows.

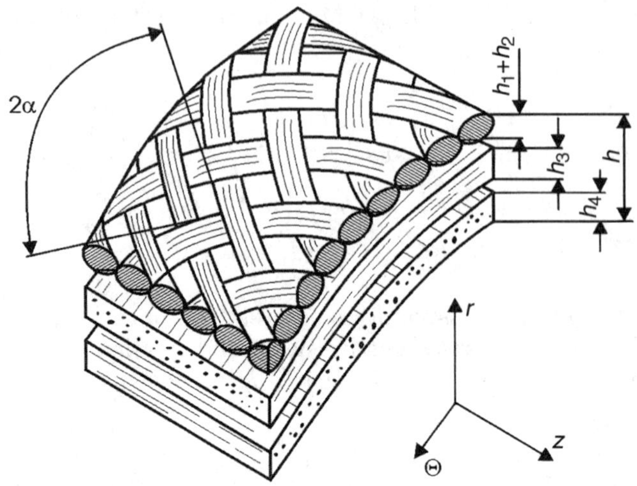

Figure 5.17 Structure of a composite material.

In presumption that the braided layer can be presented with a negligible error by two layers of equal thickness and filling degree, oriented at angle $\pm\alpha$ to the article axis, the whole packet of the material can be embodied in unidirectional layers, each having its own orientation. During the first stage, the coefficients of the pliability matrix of the unidirectional material are defined in the axes of its elastic symmetry

$$
\begin{aligned}
a_{11} &= \frac{1}{[1+(n-1)\varphi]E_{\mathrm{b}}}; \quad a_{13} = a_{12} = -\frac{\nu_{\mathrm{b}}(1-\varphi)+\nu_{\mathrm{f}}\varphi}{[1+(n-1)\varphi]E_{\mathrm{b}}} \\
a_{33} = a_{22} &= \frac{[\varphi+n(1-\varphi)][1+(n-1)\varphi]-(n\nu_{\mathrm{b}}-\nu_{\mathrm{f}})\varphi(1-\varphi)}{[1+(n-1)\varphi]E_{\mathrm{f}}} \\
a_{32} &= -\frac{[\nu_{\mathrm{f}}\varphi+n\nu_{\mathrm{b}}(1-\varphi)][1+(n-1)\varphi]-(n\nu_{\mathrm{b}}-\nu_{\mathrm{f}})\,\varphi\,(1-\varphi)}{[1+(n-1)\,\varphi]\,E_{\mathrm{f}}} \\
a_{44} &= \frac{(1-\nu_{\mathrm{f}})\varphi+n\,(1-\nu_{\mathrm{b}})(1-\varphi)}{2E_{\mathrm{f}}} \\
a_{55} = a_{66} &= \frac{(1+\nu_{\mathrm{b}})[n(1+\nu_{\mathrm{b}})(1-\varphi)+(1+\nu_{\mathrm{f}})(1+\varphi)]}{2[n(1+\nu_{\mathrm{b}})(1+\varphi)+(1+\nu_{\mathrm{f}})(1-\varphi)]\,E_{\mathrm{b}}}
\end{aligned}
\tag{5.94}
$$

where E_{f} = elasticity moduli of the fibers
E_{b} = elasticity moduli of the binder
ν_{f}, ν_{b} = Poisson ratios of the fibers and binder
φ = filling degree

After this, coefficients of the stiffness matrix of the material are found

$$
\begin{aligned}
Q_{11} &= \frac{a_{22}+a_{23}}{(a_{22}+a_{23})\,a_{11}-2a_{12}^2} \\
Q_{33} = Q_{22} &= \frac{a_{11}a_{22}-a_{12}^2}{(a_{22}^2-a_{23}^2)\,a_{11}+2a_{12}^2(a_{23}-a_{22})} \\
Q_{12} = Q_{13} &= -\frac{a_{12}}{(a_{22}+a_{23})\,a_{11}-2a_{12}^2} \\
Q_{23} &= \frac{a_{12}^2-a_{11}a_{22}}{(a_{22}^2-a_{23}^2)\,a_{11}+2a_{12}^2(a_{23}-a_{22})} \\
Q_{44} &= \frac{1}{4a_{44}}; \quad Q_{55} = Q_{66} = \frac{1}{4a_{66}}
\end{aligned}
\tag{5.95}
$$

The components of the elasticity matrix for the layers turned by angle $\pm\alpha$ to the elastic symmetry axes are defined by the following dependencies:

$$
\begin{aligned}
(\overline{Q}_{11})\alpha &= Q_{11}\cos^4\alpha + 2(Q_{12}+2Q_{66})\sin^2\alpha\cos^2\alpha + Q_{22}\sin^4\alpha \\
(\overline{Q}_{12})\alpha &= (\overline{Q}_{21})\alpha = (Q_{11}+Q_{22}-4Q_{66})\sin^2\alpha\cos^2\alpha \\
&\quad + Q_{12}(\sin^4\alpha+\cos^4\alpha) \\
(\overline{Q}_{22})\alpha &= Q_{11}\sin^4\alpha + 2(Q_{12}+2Q_{66})\sin^2\alpha\cos^2\alpha + Q_{22}\cos^4\alpha \\
(\overline{Q}_{13})\alpha &= (\overline{Q}_{31})\alpha = \sin\alpha\cos\alpha[Q_{11}\cos^2\alpha \\
&\quad - Q_{22}\sin^2\alpha + (Q_{12}+2\,Q_{66})\cos 2\alpha] \\
(\overline{Q}_{23})\alpha &= (\overline{Q}_{32})\alpha = \sin\alpha\cos\alpha[Q_{11}\sin^2\alpha \\
&\quad - Q_{22}\cos^2\alpha + (Q_{12}+2\,Q_{66})\cos 2\alpha] \\
(\overline{Q}_{33})\alpha &= (Q_{11}+Q_{22}-2Q_{12})\sin^2\alpha\cos^2\alpha + Q_{66}\cos^2 2\alpha \\
(\overline{Q}_{44})\alpha &= Q_{66}\cos^2\alpha + Q_{44}\sin^2\alpha \\
(\overline{Q}_{55})\alpha &= Q_{66}\sin^2\alpha + Q_{44}\cos^2\alpha \\
(\overline{Q}_{16})\alpha &= (Q_{11}-Q_{12}-2Q_{66})\sin\alpha\cos^3\alpha \\
&\quad + (Q_{12}-Q_{22}+2Q_{66})\sin^3\alpha\cos\alpha \\
(\overline{Q}_{26})\alpha &= (Q_{11}-Q_{12}-2Q_{66})\sin^3\alpha\cos\alpha \\
&\quad + (Q_{12}-Q_{22}+2Q_{66})\sin\alpha\cos^3\alpha \\
(\overline{Q}_{66})\alpha &= (Q_{11}+Q_{22}-2Q_{12}-2Q_{66})\sin^2\alpha\cos^2\alpha \\
&\quad + Q_{66}(\sin^4\alpha+\cos^4\alpha)
\end{aligned}
\tag{5.96}
$$

The coefficients of the elasticity matrix for the longitudinal layer can be found from relations (5.96) in presumption that $\alpha = 0$ and that, for the transversal layer, $\alpha = 90°$.

Taking into account each layer thickness, the elasticity matrix coefficients of the laminated packet are then calculated

$$A_{ij} = \frac{1}{h}\left[(\overline{Q}_{ij})_1 h_1 + (\overline{Q}_{ij})_2 h_2 + (\overline{Q}_{ij})_3 h_3 + (\overline{Q}_{ij})_4 h_4\right]; \quad i, j = 1 \ldots 6 \tag{5.97}$$

Now, based on known coefficients A_{ij}, we can find the elastic constants of the packet

$$\begin{aligned}
E_\theta &= A_{22} - \frac{A_{12}}{A_{11}}(A_{12} + A_{13}\nu_{\theta r}) + A_{23}\nu_{\theta r} \\
E_r &= A_{33} - \frac{A_{31}}{A_{11}}(A_{13} + A_{12}\nu_{r\theta}) + A_{32}\nu_{r\theta} \\
E_z &= A_{11} - \frac{A_{12}^2}{A_{22}}; \quad \nu_{z\theta} = \frac{A_{12} + A_{23}\nu_{zr}}{A_{22}} \\
\nu_{zr} &= \frac{A_{33}A_{12} - A_{31}A_{22}}{A_{22}A_{11} - A_{12}^2}; \quad \nu_{\theta r} = \frac{A_{31}A_{12} - A_{32}A_{11}}{A_{33}A_{11} - A_{31}^2} \\
\nu_{r\theta} &= \frac{A_{13}A_{12} - A_{23}A_{11}}{A_{22}A_{11} - A_{12}^2}; \quad \nu_{\theta z} = \frac{A_{12} + A_{13}\nu_{\theta r}}{A_{11}} \\
\nu_{zr} &= \frac{A_{12}\nu_{r\theta} + A_{13}}{A_{11}} \\
G_{\theta z} &= A_{66}; \quad G_{\theta r} = A_{44}; \quad G_{rz} = A_{55}
\end{aligned} \tag{5.98}$$

To define the linear thermal expansion factor of the whole laminated packet, let us first find that of a unidirectional composite[380]

$$\begin{aligned}
\alpha_1 &\approx \alpha_\mathrm{b} - (\alpha_\mathrm{b} - \alpha_\mathrm{f})\varphi\, E_\mathrm{f}/E_1 \\
\alpha_2 &\approx \alpha_\mathrm{b} + (\alpha_\mathrm{b} - \alpha_1)\nu_{21} - (\alpha_\mathrm{b} - \alpha_\mathrm{f})(1 + \nu_\mathrm{f})\frac{\nu_\mathrm{b} - \nu_{21}}{\nu_\mathrm{b} - \nu_\mathrm{f}}
\end{aligned} \tag{5.99}$$

where α_b and α_f are the linear thermal expansion factor of the binder and fiber, respectively, and E_1 and υ_{12} are the elastic constants of the unidirectional composite.

Based on the known values of the linear thermal expansion factor of the layer, let us define coefficients of thermal stresses[6] for the unidirectional layer in the axes of its elastic symmetry

$$\beta_1 = \frac{(\alpha_1 + \nu_{12}\alpha_2)E_1}{1 - \nu_{12}\nu_{21}}$$
$$\beta_2 = \frac{(\alpha_2 + \nu_{12}\alpha_1)E_2}{1 - \nu_{12}\nu_{21}} \tag{5.100}$$

Then, coefficients β_i are found in the axes turned by angle α to those of elastic symmetry of the layer

$$\beta_z = \beta_1 \cos^2\alpha + \beta_2 \sin^2\alpha$$
$$\beta_\theta = \beta_1 \sin^2\alpha + \beta_2 \cos^2\alpha \tag{5.101}$$

Thus, with allowance for thicknesses of each layer and found thermal stress coefficients of each layer, for the whole multilayered packet we obtain

$$\vec{\beta}_z = \frac{1}{h}[h_1\beta_{z1} + h_2\beta_{z2} + h_3\beta_{z3} + h_4\beta_{z4}]$$
$$\vec{\beta}_\theta = \frac{1}{h}[h_1\beta_{\theta 1} + h_2\beta_{\theta 2} + h_3\beta_{\theta 3} + h_4\beta_{\theta 4}] \tag{5.102}$$

To determine the linear thermal expansion factors of the packet,

$$\alpha_z = \frac{\vec{\beta}_z A_{\theta\theta} - \vec{\beta}_\theta A_{z\theta}}{A_{zz}A_{\theta\theta} - A_{z\theta}^2}; \quad \alpha_\theta = \frac{\vec{\beta}_\theta A_{zz} - \vec{\beta}_z A_{z\theta}}{A_{zz}A_{\theta\theta} - A_{z\theta}^2} \tag{5.103}$$

So, having found all values included in Eq. (5.93), we can determine pressure q magnitude transmitted on the mandrel from the side of the moving article. Upon using a more precise length of the mandrel in the die zone, it is possible to calculate the resistance force to motion in this part of the technological line

$$P_6 = K_D\pi \cdot d \cdot L'_D \cdot f_{fr} \tag{5.104}$$

where K_D = factor taking into consideration length of the die where dry friction is realized
L'_D = die length without entry block
f_{fr} = friction coefficient

The effect of the chief parameters on the force value (Eq. (5.106)) has been studied under the original characteristics of the material cited in Table 5.1.

Calculation results of force P_6 are presented in Fig. 5.18 depending on filling degree φ and reinforcement angle α. The analysis of results

TABLE 5.1 Principal Parameters of the Material

Name	Symbol	Value	Ref.
Cubic shrinkage ratio	ξ	2–5%	[232]
Filling degree with fibers	φ	0.4–0.6	
Linear thermal expansion factor			
Fibers	α_f	0.49×10^{-5} 1/°C	[380]
Binder	α_b	6×10^{-5} 1/°C	[380]
Elasticity modulus			
Fibers	E_f	7.4×10^4 MPa	
Binder	E_b	0.32×10^4 MPa	
Poisson's ratio			
Fibers	ν_f	0.21	
Binder	ν_b	0.38	
Reinforcement angle of the braided layer	α	±60°	
Friction coefficient	f_{fr}	0.3	[201]
Length factor of the die	K_D	0.9	
Die dimensions	r_E	11 mm	
	r_0	9.5 mm	
	L_D	500 mm	

obtained has proved that with increasing filling degree and reinforcement angle this force diminishes within 35 to 42 percent.

Changes in the resistance force are shown in Figs. 5.19 and 5.20, taking into account the relationship of the material layers. From these results, it can be concluded that introduction of a large number of

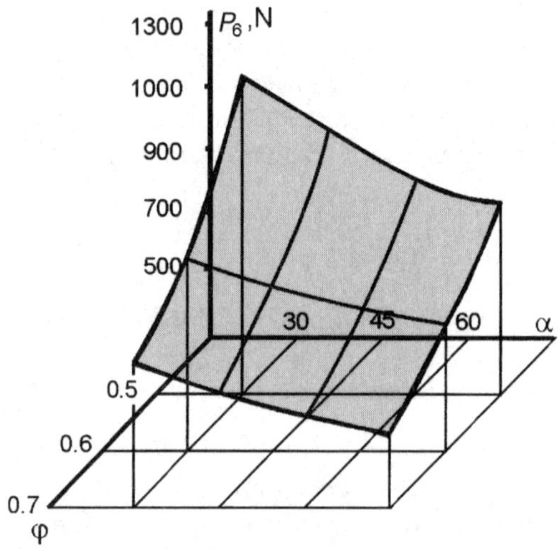

Figure 5.18 Variation of resistance force P_6 in response to reinforcement angle and filling degree of the composite.

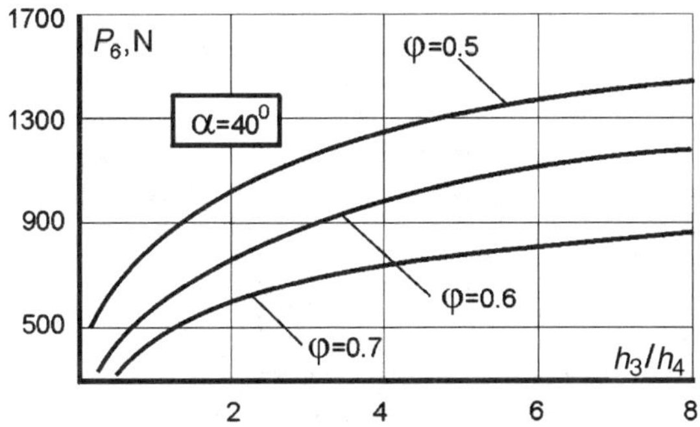

Figure 5.19 Variations in resistance force P_6 versus transversal to longitudinal layer ratio for different filling degrees of the composite.

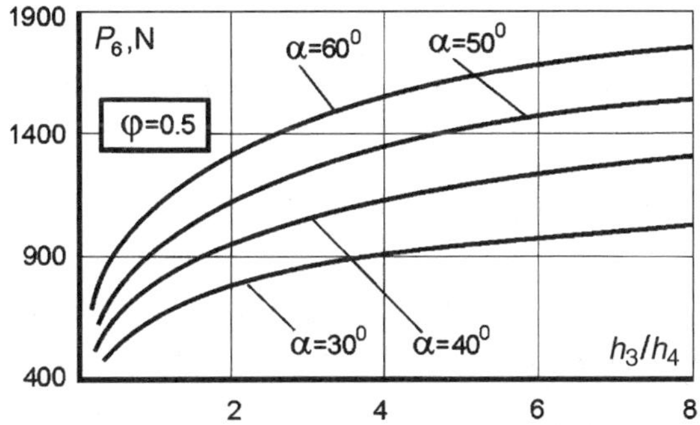

Figure 5.20 Variations in resistance force P_6 versus transversal to longitudinal layer ratio for different reinforcement angles of the composite.

transverse layers will significantly intensify the pulling force at any filling degree. However, under a high enough filling degree, this effect is exhibited less strongly. Analysis of the dependence of the pulling force on the reinforcement angle and size ratio of the layers (Fig. 5.20) has shown that the force is essentially augmented with increasing reinforcement angle. Consequently, by adjusting values φ and α, it is possible to regulate the pulling force when the size ratio of the layers in an article is to be altered.

5.6 Determination of the Total Pulling Force

The relations described above can serve the base for estimating forces in certain zones of the technological line and determining each zone length.

To calculate the resistance force in the rest zones one should remember that only sliding of a semi-finished product takes place in them

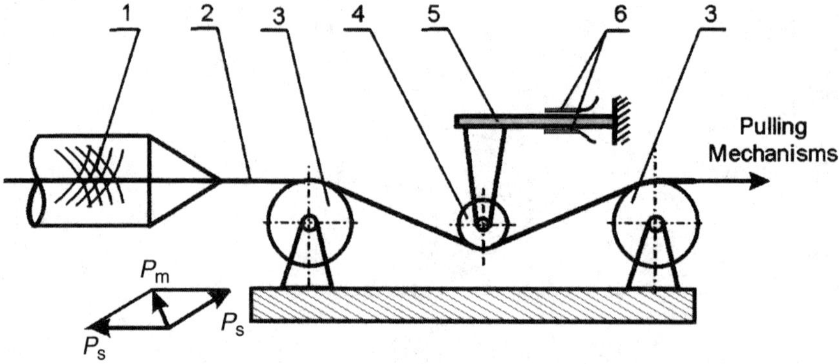

Figure 5.21 Diagram of a device for measuring pulling force: (1) article; (2) pulling rope; (3) immovable blocks; (4) running block; (5) gauge beam; (6) strain gauge; P_{m}, measured force, P_{S}, total pulling force.

without any additional depression. Hence, the total force in these zones will be

$$P_1 + P_3 + P_8 = p_0 \pi d(L_{\mathrm{M}} - L_2 - L_4 - L_6 - L_7) \qquad (5.105)$$

where L_{M} is the full length of the mandrel.

As can be seen, to calculate the total pulling force its constituents P_2 (Eq. (5.25)), P_4 (Eq. (5.50)), P_5 (Eq. (5.66)), and P_6 (Eq. (5.106)) must be found first.

To estimate the results obtained, experimental investigations have been carried out. With this aim, a special gagging device for measuring actual pulling force was elaborated and fabricated (see Fig. 5.21).

The force was measured by a gauge beam loaded by a roller found in load-carrying contact with a leader rope of the technological line. The device was calibrated using a testing machine with a rope fixed in claws, to which the gauging device was attached. Deformation of the gauge beam was estimated by the corresponding strain gauges. The pulling force was measured five to seven times within a section of stable motion of the article, upon which the measurement results for two strain gauges were averaged.

The derived experimental and theoretical results for pulling 19 mm in outer diameter rod-like articles are presented in Tables 5.2 and 5.3. The initial parameters were as follows: tension stress of the longitudinal reinforcement, 10 N/braid; tension stress of the winding transversal reinforcement, 10 N; load on each spindle of the braider, 1.5 N. Structural characteristics of the material are the following: number of longitudinal glass braids (2,500 tex), 64; thickness of the longitudinal layer, 0.65 mm; thickness of the annular layer, 0.3 mm; total thickness of the rod wall, 1.5 mm; reinforcement angle of the braided layer, $\pm 60°$.

TABLE 5.2 Force to Throughput Rate ($\phi = 0.5$) Dependence

	Pulling force P_S (N)	
Throughput rate V_S (m/min)	Theoretical value	Experimental value
0.2	5110	5220
0.3	4560	5050
0.4	4250	4050
0.5	3820	3610
0.6	3740	3570

The analysis of the data presented has proved that, with increasing throughput rate from 0.2 to 0.6 m/min, the pulling force value drops by 32 percent (according to theoretical calculations, by 27 percent). With incrementing the filling degree from 0.37 to 0.61, reached by changing the amount of longitudinal reinforcement or by introduction of greater number of threads in the transversal winding, the pulling force is augmented by 23.5 percent (theoretical, by 26 percent). It is evident that the deviation of theoretical data from experimental is not greater than 11 percent.

The pulling force variation has been studied in the course of leading part of the moistened article motion along the forming channel of the manufacturing line. The derived results are presented in Fig. 5.22 as being dependent on the pulling time. The pulling force is seen to grow with motion of the moistened part, reaching a maximum value while in the die. With increasing velocity, the force goes down.

As indicated earlier, the temperature in the die greatly influences the value of the pulling force. Experimental evidence on the die temperature effect upon the total pulling force under various velocities is presented in Fig. 5.23. It is difficult to speak unambiguously, based on these data, about the choice of the velocity and temperature regimes aimed at minimizing the pulling force. Some other parameters, including the drawn article cross-section, are to be taken into account. Variations of the pulling force in this case are illustrated in Fig. 5.24.

From experimental results, it can be concluded that the proposed mathematical model aimed at calculation of one of the most important

TABLE 5.3 Pulling Force Variation in Response to Article Filling Degree ($V_S = 0.4$)

	Pulling force P_S(N)	
Filling degree φ	Theoretical value	Experimental value
0.37	3930	3750
0.50	4250	4050
0.61	4950	4630

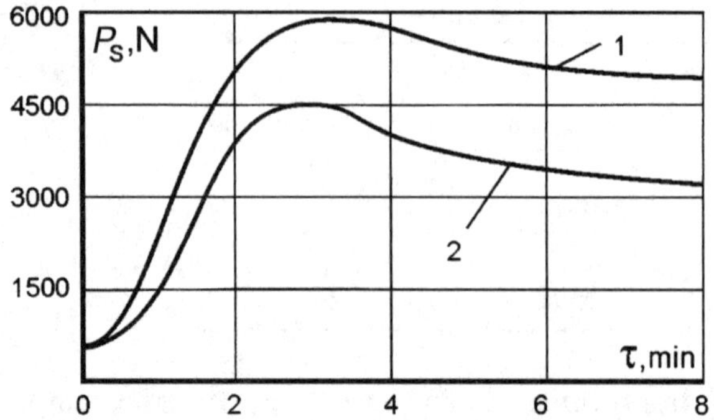

Figure 5.22 Pulling force variation at the initial stage of article motion in the technological line: (1) velocity, V_S = 0.4 m/min; (2) velocity, V_S = 0.6 m/min.

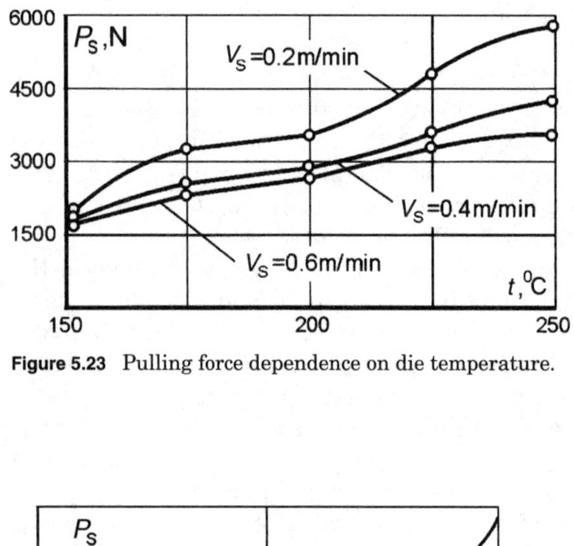

Figure 5.23 Pulling force dependence on die temperature.

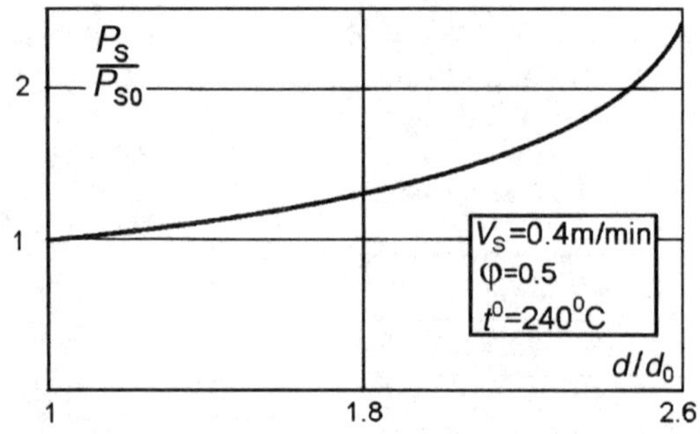

Figure 5.24 Pulling force variation with altered cross-section.

process parameters—the pulling force—quite adequately describes the procedure and is able to forecast the pulling force value as a function of other principal structural process characteristics.

5.7 Conclusions

1. Analyses of the braiding–pultrusion process have shown that one of its crucial parameters is the pulling force, its magnitude being an aggregate of resistance forces arising in separate zones of the technological line.
2. Theoretical and experimental studies of binder squeezing during braiding of the reinforcing layers have made grounds for defining the principal factors that govern resistance to motion in individual zones of the winding and braiding stages and to put forward certain recommendations on the choice of process parameters.
3. Investigations of the binder motion in the die entry have shown that the pulling force depends on geometrical parameters of the die inlet. Based on the results, the most efficient shape of the inlet that takes into account structural parameters of the article material and binder squeezing conditions has been elaborated.
4. The analysis of the stress–strain state of the article being formed with consideration of the binder shrinkage and thermal expansion of the material has made it possible to determine the resistance force to motion depending on structural parameters of the material and molding conditions in the die.
5. The conducted theoretical and experimental investigations of the resistance force to motion have established an interrelation between the total pulling force, the chief process parameters, as well as structural and geometrical characteristics of the final product material.

Chapter

6

Calculation of Force Parameters of the Molding Process of Complex-Profile Rod Elements

6.1 Main Molding Stages of Composite Complex-Profile Rod Elements

The process of molding complex-profile products by combining braiding and pultrusion has specific features that make it different from the production of unidirectional materials, i.e., from manufacture of tubular articles by braiding–pultrusion. Intricate structure, presence of a braided layer, and additional stitching impose definite limitations and specific requirements on all molding stages.

When a unidirectional profile article is pultruded, its geometrical parameters depend chiefly on the die and its shape. As for the article incorporating a braided layer, its perimeter should precisely correspond to that of the ready product and be controllable. Once formed, this layer preserves its geometrical parameters in all further operations of the molding process. As testing results have shown, violation of this condition leads to breakage of the sleeving in case of smaller perimeter and to distortion of its form and crumpling in case the perimeter is larger the required size.

Braiding over a cylindrical mandrel has been most comprehensively studied and theoretically substantiated.[3,54] During braiding, the contact sites of the annular layer, friction forces, and relative winding velocities of the braids are to be maintained constant. The contact sites of the reinforcement with the mandrel also remain invariable in order to keep the reinforcement angle and pressure of prestressed braids of the winding reinforcement constant.

This is why during preliminary molding the major tasks pursued are to shape a packet of the material of a given structure, compact it under certain pressure, remove binder excess from the longitudinal reinforcement, and obtain the needed cylindrical configuration of the semifinished product ("semi").

At the second stage, the winding reinforcement braids are packed to form the required braided structure and attain the sought reinforcement angle of the sleeving. Under tension forces of the winding braids, the semi is further molded and the winding reinforcement is partially impregnated with the binder.

During the third stage, to obtain the required perimeter of the semifinished product, it is calibrated immediately after the braiding layer has been laid, otherwise the sleeving might be crumpled and corrugated because of elongation of the winding reinforcement in the axial direction under the pressure effect. Operations in the second and third stages are usually joined together, with which aim the calibrating die is installed in the section where the one-sheet hyperboloid of revolution of the free braiding surface intersects the cylindrical surface of the semi. If the die is installed ahead of the section, the braids would experience lateral friction forces, evoking twisting of the braids and distortion of the braiding layer structure.

The next stage of the force treatment is molding of the profile article post. The semi is pulled off the mandrel and is compacted where the post is to be formed. The post is then separated into zones of the shoulders and the flange using special plates fixed on the mandrel. The post height is marked and separated from the shoulders with the help of stitching the junction. After these steps, the post can be finally formed.

During the final manufacturing stage, the flanges of the profile element are molded at a preset profile configuration, filling degree of the main material of the flanges, and additional impregnation of the winding braids. This process can be combined with the final molding stage of the post. This process goes hand-in-hand with the polymer binder curing and fixation of the article shape.

6.2 Pressure Estimation at Preforming a Semi of the Rod Element

The packet of the material preformed during the first stage incorporates binder-impregnated reinforcing materials of all structural elements of the profile. The problems to be solved at this stage depend on the type of materials chosen and their impregnation conditions.

In wet impregnation of all reinforcement components, the semifinished product might contain excess binder that should be removed to attain the required filling degree. The amount of binder remained ensures the needed filling degrees of each structural element, including the braiding layer. In case of wet impregnation of only a part of the reinforcement material, just this part should incorporate the amount of the binder to allow for further saturation of the braiding layer. When predried prepregs are used, the sleeving is to be impregnated individually.

In practice, the second combined method, for which a veil for molding the post and shoulders is taken in the form of a prepreg and the longitudinal reinforcement of the flanges is wet-impregnated, is more convenient.

The filling degree of all constituent materials by a fibrous reinforcement is specified by their type.

The volume filling degree of a unidirectional composite and braids is commonly $\varphi \approx 0.60$–0.65, while that for veils is 0.4–0.5. The braiding layer filling can be found by the formula

$$\varphi = \frac{\varphi_1 t_b}{\frac{2\Pi}{m}\sin\alpha + t_b} \tag{6.1}$$

where t_b is the width of the winding braids; Π is the perimeter of the complex-profile article; m is the number of spindles of the braider; and α is the reinforcement angle of the braiding layer.

The value of φ is usually rather low; even in conditions of dense packing of the braids over the article surface, it is half that of the braids.

To ensure a given filling degree of each structural element of the profile, it is calculated with allowances for further processing stages as well as for preset geometrical parameters of the die and mandrel.

The excess binder area in the semisection needed for further impregnation of the braiding layer can be calculated as

$$F^* = F(1 - \varphi) \tag{6.2}$$

where F^* is the total area of the braiding layer.

At the second stage of preforming, there arises a radial pressure constituent under the action of tension forces of the winding braids T leading to molding of the main material (Fig. 6.1)

$$p_0 = \frac{T \cdot \sin^2\alpha}{t_b D_d} \tag{6.3}$$

where D_d is the diameter of the forming die.

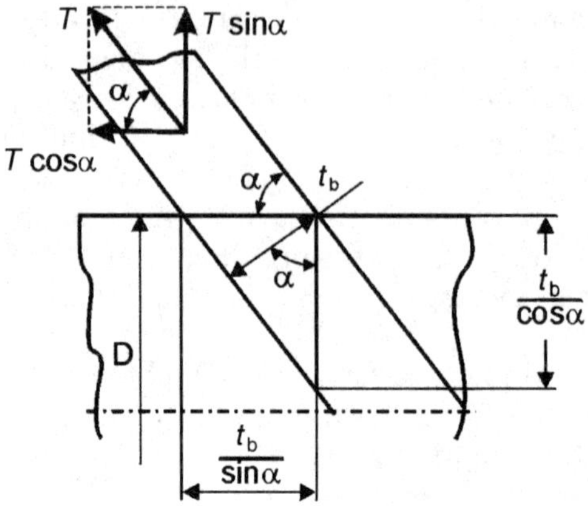

Figure 6.1 Distribution of tension forces at braiding.

This pressure generates the redistribution processes of the binder inside the semi structure, that is, the pressure displaces the viscous liquid in the porous medium. The duration of this process is restricted by the rate of an elementary unit formation of the braiding layer, i.e. a cell equal to

$$t = \frac{2}{mn_0} = \frac{t_{\mathrm{b}}}{\pi D_{\mathrm{d}} n_0 \cos\alpha} \tag{6.4}$$

where n_0 is the rotation velocity of braider's spindles.

The viscosity of the binder η governs the binder redistribution intensity, which can be regulated within certain limits by varying either the processing temperature or the solvent concentration.

In presumption of a laminar flow of the liquid binder at a constant temperature, we can write Darcy's relation analogous to Eq. (5.6) as follows:[55,194,216]

$$\overline{v} = \frac{dL}{dt} = \frac{k_{\mathrm{p}}\Delta p}{\eta L} \tag{6.5}$$

where $\overline{v}$ is the mean velocity of the liquid phase and where the permeability factor k_{p} can be found from Eq. (5.7).

Displacements of the binder flows under molding pressure $\Delta p = p_0$ are shown in Fig. 6.2.

The total amount of the binder having passed through the surfaces under study is defined as a sum of the flows.

$$Q = Q_x + Q_{\mathrm{r}} \tag{6.6}$$

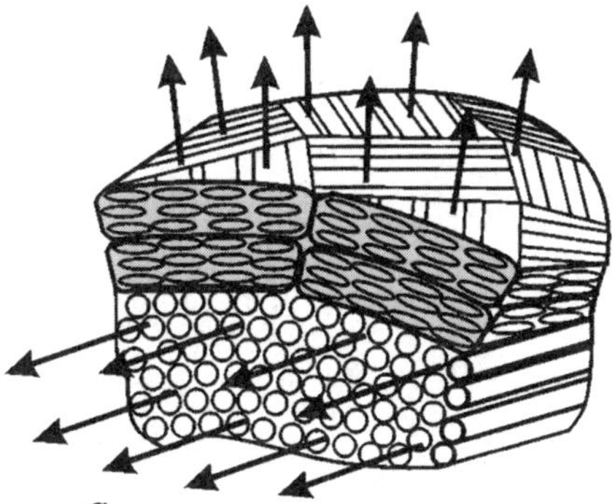

Figure 6.2 Spreading of binder flows during braiding.

The binder flow passing in the axial direction through a unit surface of the considered element to distance L is determined by the formula

$$q_x = (\overline{\nu} - V_S) \cdot S_{xb} \tag{6.7}$$

where S_{xb} is the area occupied by the binder on the surface through which the flow passes.

In the longitudinal direction the binder should not be squeezed so that $q_x = 0$ and

$$L \le V_S t \tag{6.8}$$

Considering that $S_{xb} \neq 0$, we obtain $\overline{\nu} = V_S$.

Then, from Eq. (6.5) and taking into account Eqs. (6.4) and (6.8), one can obtain the rated value of the molding pressure that meets the preset conditions.

$$p_0 \le \frac{2V_S^2 \eta}{k_p^L m n_0} \tag{6.9}$$

This is the pressure that is created by tension forces of two winding braids at forming the braiding layer cell, where each tension force can be defined by substitution of Eq. (6.9) in Eq. (6.1)

$$T \le \frac{V_S^2 \eta\, D_d t_b}{k_p m n_0 \sin^2 \alpha} \tag{6.10}$$

The value of k_p^L can be obtained when it is supposed that the fibers of the reinforcing material are deflected negligibly

$$k_p^L = \frac{(1 - \varphi_1)^3 d_f^2}{16 k_6 \varphi_1^2} \tag{6.11}$$

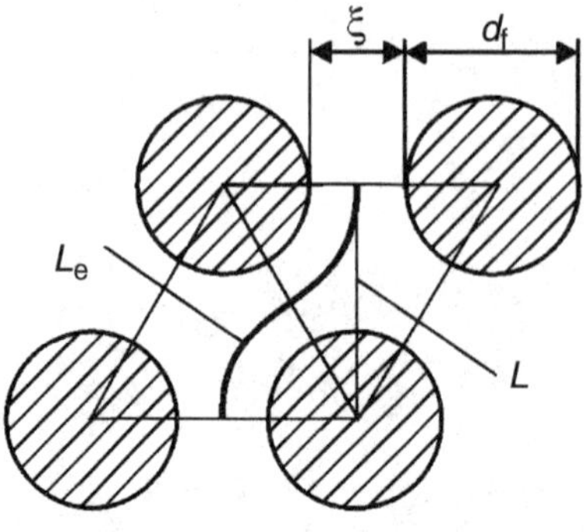

Figure 6.3 Migration path of the binder during braiding in the radial direction.

The binder flow in radial direction in the sleeving zone moves through the side surface of the braiding layer, and, therefore, channel sizes are similar to the distances between fibers ξ (Fig. 6.3). In this case, the permeability factor $k_{\mathrm{p}}^{\mathrm{r}}$ can be determined by relation (5.7) as well, in which the tortuosity factor is $L/L_{\mathrm{e}} = 0.827$

$$k_{\mathrm{p}}^{\mathrm{r}} = 0.827\, k_{\mathrm{p}}^{\mathrm{L}} \approx \frac{(1-\varphi_1)^3 d_{\mathrm{f}}^2}{20\, k_6\, \varphi_1^2} \tag{6.12}$$

The depth of saturation of the braiding layer and time are interrelated by the following equation:[218]

$$L_{\mathrm{r}} = \sqrt{2\frac{k_{\mathrm{p}}^{\mathrm{r}}}{\eta} p_0 t} \tag{6.13}$$

If L_{r} is larger the braiding layer thickness δ_{b}, the binder will then be squeezed onto the braiding layer surface and squashed, which is undesirable. At $L_{\mathrm{r}} < \delta_{\mathrm{b}}$, the sleeve will stay dry.

Proceeding from condition above, the required molding pressure can then be calculated by the formula

$$p_0 < \frac{\delta_{\mathrm{b}}^2 \eta}{2\, k_{\mathrm{p}}^{\mathrm{r}} t} \tag{6.14}$$

Hence, the rated tension force value of the winding braids, taking Eqs. (6.3) and (6.4) into account, will be

$$T < \frac{\delta_{\mathrm{b}}^2 \eta\, D_{\mathrm{d}} t_{\mathrm{b}} m n_0}{8\, k_{\mathrm{p}}^{\mathrm{r}} \sin^2 \alpha} \tag{6.15}$$

As far as both conditions are to be met at manufacture of complex-profile articles, to select the tension value of the braids, formulas (6.10) and (6.15) should both be used and the lesser of results should be chosen.

Once the tension force T is found, the rated impregnation depth of the braiding layer $L_r^C < \delta_b$ can be calculated based on Eq. (6.15)

$$L_r^C = \sqrt{\frac{8k_p^r T}{\eta D_d t_b m n_0}} \cdot \sin\alpha \tag{6.16}$$

Hence, the volume of the binder having passed through a unit area of the article surface can be calculated by the relation

$$Q_r' = (1 - \varphi_1) \cdot L_r^C \tag{6.17}$$

The full volume of the binder spent on impregnation of the braiding layer can be found when this value is multiplied by the impregnation area. If the parameters of the mandrel and those of the die have been calculated correctly, then the impregnation area of a unit length of the semi should be numerically equal to the article perimeter without the braiding layer. So far,

$$Q_r^* = (1 - \varphi_1) \cdot L_r^C \Pi^* \tag{6.18}$$

The same equation can also be presented as

$$Q_r^* = (1 - \varphi) \cdot L_r^R \Pi^* \tag{6.19}$$

The actual impregnation depth of the braiding layer L_r^R turns out to be less than the rated one because a part of the binder is used to fill the free space of the braiding cell. Proceeding from a condition of the equality of binder volumes, it can be calculated as

$$L_r^R = L_r^C \cdot \frac{(1 - \varphi_1)}{(1 - \varphi_{bl})} \tag{6.20}$$

After the braiding layer has been wound over the semifinished product, it is fed in a calibrating die to adjust its perimeter. Its structural parameters are in line with a compacted state of the braiding layer, so the calibrating die will not essentially change the impregnation characteristics.

The actual impregnation depth can be also determined experimentally. To estimate this parameter, glass-reinforced plastic samples were fabricated with the braided layer having different tension force values. To eliminate various side effects induced by changes in the binder viscosity in response to temperature variations and to ensure the needed polymerization level at 20°C, a cold-set epoxy binder with a resin-to-hardener ratio of 9/1 has been used. Because the binder polymerization process is exothermal, the binder has been prepared in small portions.

Analogous investigations were carried out to study the impregnation depth dependence on the binder viscosity. The tension force of the winding braids was set to 10 N. The viscosity was varied by diluting the binder with an alcohol–acetone mixture. Because it is difficult to directly estimate the viscosities of the compound, the diluted resin (η_b), and the hardener (η_h), they were estimated by a viscosimeter, and the sought value was calculated following the rule of mixtures

$$\eta = 0.1\eta_h + 0.9\eta_b \tag{6.21}$$

Attempts to estimate the impregnation depth by microstructural analyses have resulted in a considerable scatter of results. This is because the impregnation depth is a statistical parameter that depends on a number of factors, which are partly connected with imperfections of the material structure. Thus an integral estimate has been employed to obtain L_r^R experimentally, which consists of determining the weight content of the binder in the impregnated braiding layer of the sample under study as it is burned off. Based on the results, the integral values of the impregnated layer thickness are then calculated as

$$L_r' = \frac{G_b}{\gamma_b F(1 - \varphi_{bl})} \tag{6.22}$$

where γ_b and G_b are the specific and actual weights of the binder in the sample, respectively, and F is the surface area of the sample.

Theoretical dependencies and experimental results presented in Figs. 6.4 and 6.5 characterize the effect of tension forces of the braids and viscosity of the binder used on the braiding layer impregnation depth.

As analysis of the charts has proved, the experimental results comply well with calculations. Consequently, the derived theoretical relations can be used in designing process parameters.

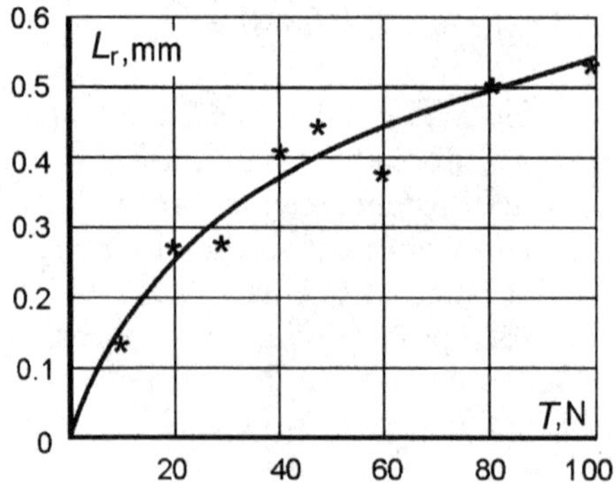

Figure 6.4 Effect of tension forces on impregnation depth of the braiding layer.

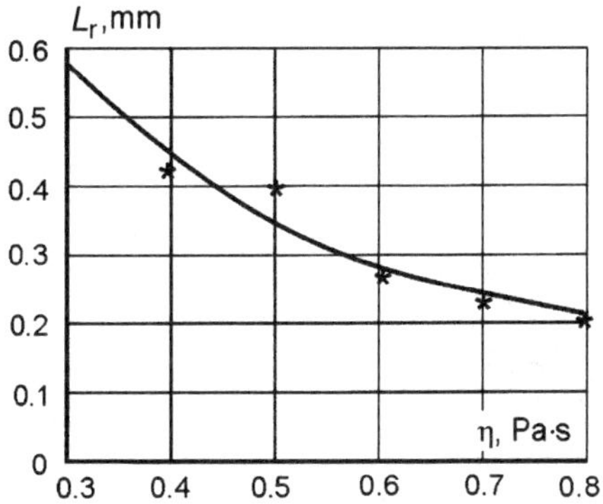

Figure 6.5 Effect of binder viscosity on impregnation depth of the braiding layer.

6.3 Study of Impregnation During Final Molding of the Complex-Profile Article

The article profile can be finally shaped upon its calibration unless it is necessary to mold a post and flanges and the article requires stitching.

The process of shaping the post is subdivided into two stages, namely, preforming and final molding.

In the first stage, a flat profile wall is shaped from a cylindrical semifinished product. For this purpose, the problems of centering the profile, separation of the material of the post and flanges, partial compaction of the post material, and fixation of its boundaries by stitching are to be solved. These are the design problems, and they should be solved at the stage of designing the rigging and equipment. Compaction of the material may solve only the problem of joining its parts situated on opposite sides of the semi, so pressure should not be too high because the fibers are to freely move during stitching.

The maximal pressure in this case is usually limited to that arising at deformation of prestrained elements of the material forming the post. It is governed by tension forces of these elements and curvature radius of compacted surfaces R_y (Fig. 6.6)

$$p_\mathrm{r} = \frac{P_\mathrm{em} F_\mathrm{r}}{2 F h_\mathrm{r} R_y} \tag{6.23}$$

where P_em is the tension force of the material elements, F and F_r are the cross-sectional area of the profile and post, respectively, and h_r is the profile post height.

In case of wet impregnation of the post material, partial impregnation of the braiding layer will occur. The impregnation mechanism considered above will be realized if the pressure p_r obeys the relation

$$p_r \le \frac{\eta V_\mathrm{S} R_y}{k_\mathrm{p}^\mathrm{r}} \arctan\left(\frac{D_\mathrm{c} - b_r}{2L_y}\right) \tag{6.24}$$

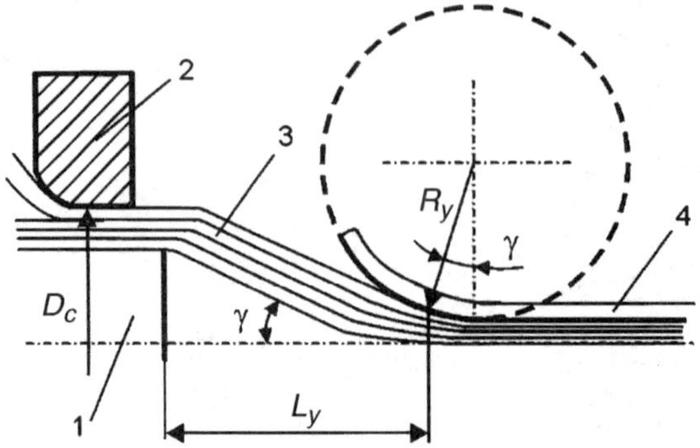

Figure 6.6 Initial molding of the profile post: (1) mandrel; (2) calibrating die; (3) molded material; (4) molding device.

where L_y is a structural parameter equal to a distance from the point of the semi leaving the mandrel to its contact with the compacting device (Fig. 6.6), b_r is the post width, and D_c is the open-flow size of the calibrating die.

When a prepreg is employed, the binder is not squeezed at this stage. The post of the profile article is shaped completely during final molding. This operation can be performed either jointly with molding of the flanges or separately. In the former case, the profile is shaped in a special forming die whose dimensions ensure the needed geometrical parameters of the final product. In this die, the binder is squeezed to the required amount and is redistributed over the profile section, the braiding layer being additionally impregnated and the article cured.

In the latter case, geometrical parameters of the post are preset prior to formation of the flanges by passing it between two forming plates or rollers heated to a temperature of a given viscosity of the binder. Two advantages of this method are lowered resistance to pulling and the possibility of joining this operation with stitching. It is preferable to use this method for large-dimension products. During this stage, the forming die performs molding of the profile flanges, calibration of the post size, additional impregnation of the braiding layer, and article curing.

In the course of the mentioned procedures, significant pressure may arise, in which case Eq. (6.24) is not obeyed. As a consequence, the binder moves in longitudinally and is squeezed transversally not only in molding devices but also outside of their range.

Let us define pressure gradient along the molded post prior to the molding front (Fig. 6.7). For a mathematical model of the processes taking place at post molding, the binder, as in the preceding chapter, is taken as a viscous, incompressible liquid that obeys Darcy's law at

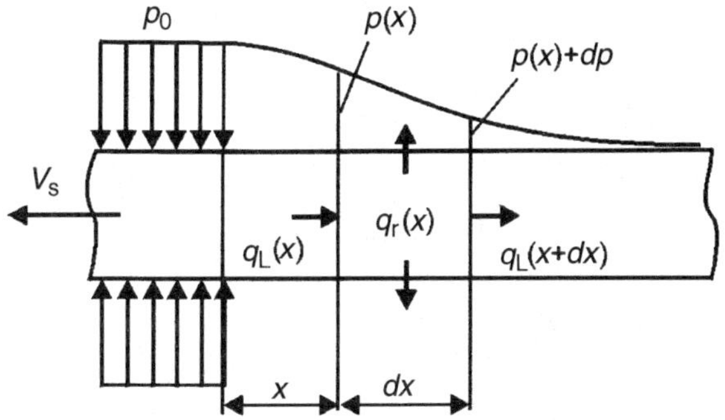

Figure 6.7 Pressure and binder flow distribution during final molding of the profile article post.

squeezing. The liquid binder flow will be considered separately from the stress state of the stiff reinforcement.

If to keep to the calculation scheme presented in Sec. 5.4, the flow of the binder to be passing through the element cross-section of length dx and coordinate x in the longitudinal direction can be found by Eq. (5.67)

$$q_{\rm L}(x) = \frac{k_{\rm p}^{\rm L} p(x)}{\eta x} \cdot A_{\rm L} \tag{6.25}$$

and at distance $(x + dx)$

$$q_{\rm L}(x + dx) = \frac{k_{\rm p}^{\rm L}[p(x) + dp]}{\eta(x + dx)} \cdot A_{\rm L} \approx \frac{k_{\rm np}^{\rm L} p(x) A_{\rm L}}{\eta x} + \frac{k_{\rm np}^{\rm L} A_{\rm L} dp}{\eta x} \tag{6.26}$$

The area through which the binder flow passes lengthwise is found in terms of that of the element being molded:

$$A_{\rm L} = (1 - \varphi_1) \cdot F_{\rm r} = (1 - \varphi_1) \cdot b_{\rm r} h_{\rm r} \tag{6.27}$$

A part of the binder is squeezed from the element volume in the radial direction. Considering that the dependence $p(x)$ is linear in a section dx, it is justified to write

$$q_{\rm r}(x + dx) = \frac{2k_{\rm p}^{\rm r}\left[p(x) + \frac{dp}{2}\right]}{\eta b_{\rm r}} \cdot S_{\rm r} \tag{6.28}$$

Permeability in the radial direction will be calculated by the relation

$$S_{\rm r} = n_{\rm r}\xi^2 \tag{6.29}$$

where $n_{\rm r}$ is the number of filtration channels and ξ is the distance between neighboring fibers.

If the fibers are presumed to be arranged in a honeycomb structure in the post, the number of channels is

$$n_r = \frac{2h_r dx}{(d+\xi)^2} \tag{6.30}$$

The distance ξ can be found with the filling degree φ_1 presented as a product[55]

$$\varphi_1 = \psi_1 \cdot \psi_2$$

where ψ_1 is a structural characteristic governed by the fiber packing type and is equal to the filling degree with maximum dense packing. Optimum in radial direction is considered to be hexagonal, which allows for the largest gap between fibers.

In this case, $\psi_1 \approx 0.907$ and ψ_2 is a geometrical characteristic that defines the relation of distances between fibers ξ to filling degree (Fig. 6.3)

$$\psi_2 = \frac{1}{(1+\xi/d_f)}$$

Hence, distance between fibers can be calculated by the formula

$$\xi = d_f\left(\sqrt{\frac{0.907}{\varphi_1}} - 1\right) \tag{6.31}$$

If we substitute Eqs. (6.30) and (6.31) in Eq. (6.29), we obtain

$$S_r \approx 2.2 h_r \varphi_1 \left(\sqrt{\frac{0.907}{\varphi_1}} - 1\right)^2 dx \tag{6.32}$$

The equation for the balance of flows based on Eqs. (6.25), (6.26), and (6.28) is

$$\frac{dp}{p(x)} = \frac{3\,\varphi_1 \left(\sqrt{\frac{0.907}{\varphi_1}} - 1\right)^2 x dx}{b_r^2(1-\varphi_1)} = \beta x dx \tag{6.33}$$

where

$$\beta = \frac{3\,\varphi_1 \left(\sqrt{\frac{0.907}{\varphi_1}} - 1\right)^2}{b_r^2(1-\varphi_1)} \tag{6.34}$$

As shown in Chap. 5, solution of the relation above is of the form

$$\ln p(x) = -\beta x^2 + \ln C \quad \text{or} \quad p(x) = C \cdot e^{-\beta x^2}$$

The constant C can be found from analogous original conditions at $x = 0, \Rightarrow C = p_r$.

Thus, the law of pressure variation is described in this case by an exponential dependence

$$p(x) = p_r \cdot e^{-\beta x^2} \tag{6.35}$$

The volume of the binder removed at post molding per unit time should evidently be equal to the total volume of the binder passing through its surface.

$$Q_r = \int_0^{\infty} \frac{k_{\mathrm{p}}^{\mathrm{L}} F_r (1 - \varphi_1)}{\eta} \cdot \beta p_{\mathrm{r}} e^{-\beta x^2} dx \tag{6.36}$$

By making use of $\int_0^{\infty} e^{-\beta x^2} dx = 0.5\sqrt{\pi/\beta}$, after substitutions we have

$$Q_r = \frac{\sqrt{3\pi}\, k_{\mathrm{p}}^{\mathrm{L}}}{2\eta} h_r p_r \sqrt{\varphi_1 (1 - \varphi_1)} \left(\sqrt{\frac{0.907}{\varphi_1}} - 1 \right) \tag{6.37}$$

The binder area in the article cross-section that will guarantee pressure p_r is obtained if the removed volume is divided by the pulling velocity V_{S}.

If the amount of the binder in the semi structure is enough for full saturation of the considered area of the braiding layer, its surplus area can be calculated as the difference between the total area found by Eq. (6.2) and that of the binder used calculated by Eq. (6.19) as follows:

$$\begin{aligned} F_0^{\mathrm{c}} &= 2(1 - \varphi)\, \delta_{\mathrm{b}} h_r - 2(1 - \varphi_{\mathrm{bl}})\, L_r^{\mathrm{R}} h_r \\ &= 2\, h_r (1 - \varphi)\big(\delta_{\mathrm{b}} - L_r^{\mathrm{R}}\big) \end{aligned} \tag{6.38}$$

Consequently, molding pressure arising in the molding device will then be

$$p_r = \frac{1.3\, \eta V_{\mathrm{S}} (1 - \varphi_{\mathrm{bl}}) (\delta_{\mathrm{b}} - L_r^{\mathrm{R}})}{k_{\mathrm{p}}^{\mathrm{L}} \sqrt{\varphi_1 (1 - \varphi_1)} \left(\sqrt{\frac{0.907}{\varphi_1}} - 1 \right)} \tag{6.39}$$

As it follows from the analysis, the formulas obtained can be used to find pressures in the forming device or the die during molding of any element of the profile. In this case, the variable parameters are the permeability factor in the longitudinal direction, which is a function of structural characteristics of the main material of the profile article; binder viscosity, which is governed by the treatment temperature; and the relative saturation depth of the braiding layer attained at preliminary stages of force treatment of the semi. Structural analysis of

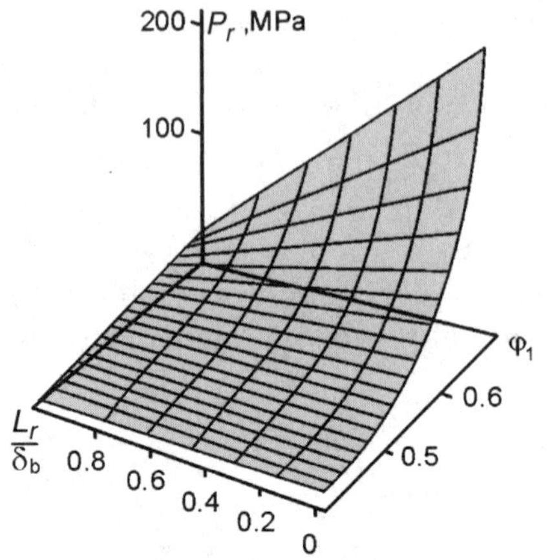

Figure 6.8 Pressure dependence on filling degree and preimpregnation of the braiding layer at final molding of the profile.

the formulas has shown that the pressure generated in the material is linearly dependent on binder viscosity, pulling velocity, and degree of preimpregnation of the braiding layer. Simultaneously, it is a complex function of the filling degree because it governs both the permeability factor and filling degree.

A dependence of the maximum pressure upon filling degree and relative preimpregnation depth of the sleeving is shown in Fig. 6.8. An increasing filling degree is seen to augment pressure that may reach a magnitude commensurable to fiber strength at shear.

6.4 Design of Resistance Forces at Various Molding Stages of Complex-Profile Articles

Resistance forces to motion of the reinforcement, the semi, and the article as a whole during the different process stages govern the pulling force and are most important parameters of the process.[205] Their increase above certain limits may result in failure of the reinforcing material.

In general, except for the pretension forces of the fibrous reinforcement, the forces mentioned above can be lowered by introduction of special anti-stick additives into the binder, control over temperature and pressure regimes, and so on.[207,220]

The total pulling force of a complex-profile article specified by a pulling mechanism can be presented as a sum of resistances to motion at various stages of the article being formed (Fig. 6.9).

All operating resistance forces can be conditionally subdivided into three main types:

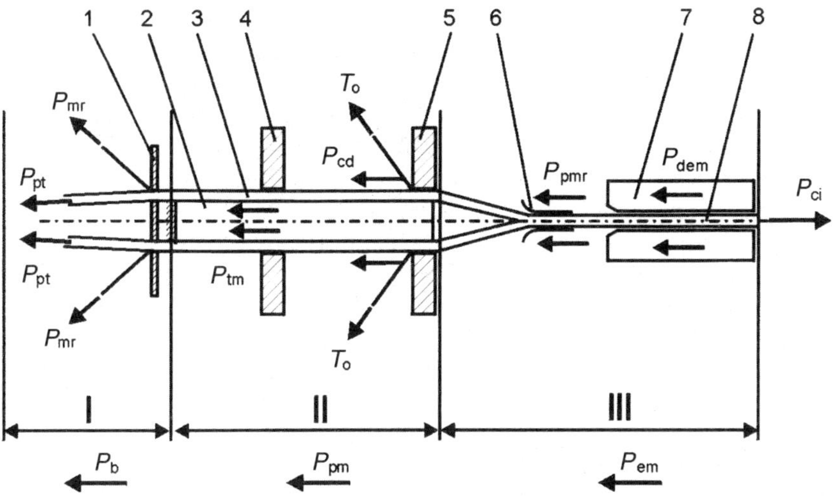

Figure 6.9 Distribution of resistance forces during manufacture of complex-profile composite ware: (1) distribution device; (2) mandrel; (3) molded material; (4) premolding die; (5) calibrating die; (6) premolding device of the profile post; (7) final molding die; (8) article being shaped.

1. Pretension forces P_b consisting of
   - Pretension forces of the main reinforcement braids, P_{pt}
   - Forces generated by the impregnation and tension devices, P_{ti}
   - Tension forces of the auxiliary material of the post, P_{mr}
2. Premolding forces P_{pm} consisting of
   - Resistance forces in the die at premolding, P_{dpm}
   - Resistance forces to motion over the mandrel, P_{tm}
   - Tension forces of the winding reinforcement braids, P_{ba}
   - Resistance forces to calibrating die, P_{cd}
3. Forces of final molding, including
   - Resistance forces of the premolding device of the post P_{pmr}
   - Resistance forces of the final molding die, P_{dem}

It should be noted that each resistance force considered has, in turn, a complex structure. The first three resistance forces P_b are the process tension forces the make the main reinforcement fibers linear during all further molding stages.

Without certain preliminary tension of the braids, the strainers in the equipment cannot work. When they are tightened, the braids bend around the rollers of the impregnator, bringing about both axial and radial constituents of the pressure that improve impregnation quality.[195] This preliminary tension is usually chosen experimentally, being about 10–15 N/braid for glass and organic fibers and 1–3 N/braid for the carbon-fiber based braids.

When they reach the impregnation bath, the braids are wetted by the binder heated to a definite temperature to guarantee the needed viscosity and impregnation quality. When they are bent on the rollers of the impregnation line, the braids are slowed down slightly. As a result, an additional tension force arises whose value depends upon the device design. The total tension force in the impregnation bath outlet is usually not high, reaching 3–5 N/braid for carbon braids.

The tension force of the auxiliary reinforcement of the post is chosen similarly to above described. Preimpregnated cloth strips are often used for the post, whose tension force should in addition to other tasks maintain the strip linear when the adhesive film is peeled off its surface. The value of this tension force is regulated by lightly braking the reel with the material by a special band brake or by some other method.

As the main reinforcement material approaches the molding zone, it is spread over the mandrel and fed to the die for preliminary molding. In this case, the resistance force is composed of the friction and adhesive forces and can be calculated analogously to Eq. (5.25) in terms of the generalized law of friction[201]

$$P_{\mathrm{dpm}} = N \cdot f_{\mathrm{fr}} + \tau_{\mathrm{at}} F_{\mathrm{a}} \tag{6.40}$$

where τ_{at} is the adhesive tangential stresses generated at relative shear of contacting surfaces (Eq. (5.27)), N is the normal constituent of pressure on the mandrel, and F_a is the area of adhesion of the whole contour of the complex-profile article.

Excess binder from the braids is squeezed out prior to the braid entering the die, and the pressure near the die inlet rises, so fibers can be subjected to free deformation. Because the area of squeezing is rather large and the pressure contribution in the total resistance force does not surpass 10 percent, the friction forces in the die can be neglected.

Experimental investigations of the relationship of adhesive stresses and pulling velocity of a complex profile article under various temperature regimes have shown that the stresses diminish at all temperatures as the velocity increases. These results match up data in Chap. 5 (Fig. 5.3).

Adhesive forces are effective over the entire surface of the mandrel and premolding die. Therefore, the sum of the mandrel and die areas contacting the impregnated semi can be substituted in Eq. (6.40)

$$P_{\mathrm{tm}} = \tau_{\mathrm{at}}(\pi D_{\mathrm{pd}} L_{\mathrm{pd}} + \Pi_{\mathrm{m}} L_{\mathrm{m}}) \tag{6.41}$$

D_{pd} and L_{pd} are the diameter and the length of the premolding die, and D_m and L_m are the perimeter and length of the mandrel contacting the semi.

The resistance at molding the braiding layer is created by tension forces in the winding reinforcement and friction forces in the calibrating die. Proceeding from Eqs. (6.10) and (6.17), each mth braid of the winding reinforcement acquires some initial tension T_b that is directed at angle α to the mandrel axis. Hence, the total axial constituent of all braids will be calculated as

$$P_{ba} = m \cdot T_b \cos\alpha \tag{6.42}$$

This formula is true if the condition of locating the calibrating die in the section complying with the winding reinforcement packing is satisfied, or otherwise angle α would change and it would be necessary to add the friction forces of the braids at the die inlet to Eq. (6.42).

According to Sec. 6.2, the calibrating die operates jointly with the winding braids and calibrates the semi perimeter until it reaches the rated value. Note that pressures generated in the die are calculated with Eq. (6.3). Because the winding reinforcement has not yet been impregnated, along with adhesive forces exerted by the mandrel there appear friction forces calculated by the formula

$$P_{cd} = \frac{p_0}{2}\pi D_c L_c f_{fr} \tag{6.43}$$

where D_c and L_c are the dimensions of the calibrating die, provided that the pressure is spread linearly along the die from p_0 to 0.

Resistance forces generated in premolding devices can be subdivided into two types:

- Forces of frontal resistance $P_{pmr}{}^{R}$ brought about by changes in direction of the post material motion
- Friction forces that arise as the moving material of the post bends around the curvilinear surface of the forming element

The frontal resistance force is found by integrating axial constituents of normal stresses by the wrap angle of the forming element by the material (Fig. 6.10)

$$\begin{aligned} P_{pmr}{}^{R} &= 2\int_0^{\gamma} p_r h_r R_y \sin\alpha\, d\alpha = \int_0^{\gamma} \frac{(P_b + P_{pm})\,F_r}{F}\sin\alpha\, d\alpha \\ &= \frac{(P_b + P_{pm})\,F_r}{F}(1-\cos\gamma) \end{aligned} \tag{6.44}$$

In relation (6.44), p_r is determined by Eq. (6.23) using the supposition that pressure on the forming element is created by the post material, whose resistance forces have been acting in the previous stages.

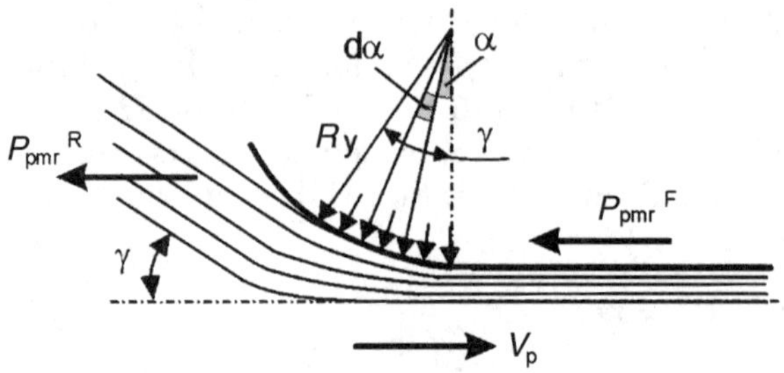

Figure 6.10 Design scheme of resistance forces at premolding of the post.

The value of γ shown in Fig. 6.6 is

$$\gamma = \arctan\left(\frac{D_c - b_r}{2L_y}\right) \tag{6.45}$$

The friction force can be calculated by Euler's formula

$$P_{pmr}{}^{F} = \frac{(P_b + P_{pm})\,F_r}{F}\cdot e^{f\cdot\gamma} \tag{6.46}$$

Hence, the total resistance force on this section is determined by a sum

$$P_{pmr} = P_{pmr}{}^{R} + P_{pmr}{}^{F} = \frac{(P_b + P_{pm})\,F_r}{F}\cdot(1 - \cos\gamma + e^{f\cdot\gamma}) \tag{6.47}$$

It is evident from Eq. (6.46) that with increasing wrap angle the constituent of the resistance, which is a function of the friction force, grows quickly, too. This is why it is expedient to use roller mechanisms to reduce the resistance when molding large-size products.

During complete molding, the resistance forces depend on three main factors, namely, frontal resistance at the inlet to the die, forces arising from binder redistribution within the material structure, and forces of the article adhesion to the forming die surface.

Frontal resistance is practically absent when the profile post is finally formed as the fibers upon molding enter the die at an angle close to zero.

The frontal resistance at flange molding can be calculated analogously to that of the post by Eq. (6.45). One should, however, remember that various layers of the flange material deviate to different angles during molding depending on the difference in coordinates of their positions before and after molding (Fig. 6.11).

The resistance force of a unit layer is then

$$dP_{fs}^{R} = \frac{(P_b + P_{pm})\,B_f\,dy}{F}(1 - \cos\gamma) \tag{6.48}$$

where $dy = R^{*}\cdot d\gamma$ and B_f is the flange width.

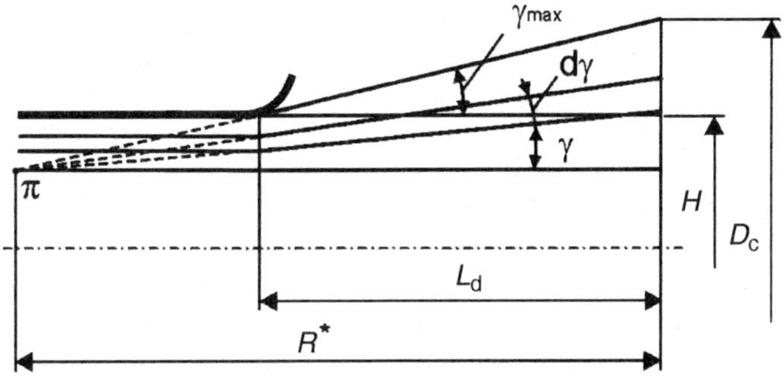

Figure 6.11 Deflection chart of various layers of the flange material at its final molding.

In addition, R^* can be found from the relation

$$\frac{D_c - h_r}{2R^*} = \tan \gamma_{\max} = \frac{D_c - H}{2L_d} \tag{6.49}$$

where H is the profile height; thus

$$R^* = \frac{D_c - h_r}{D_c - H} \cdot L_d$$

After substitutions in Eqs. (6.48) and (6.50) and integration, we obtain

$$P_{fs}^{R} = \frac{(P_b + P_{pm})\, B_f L_d (D_c - h_r)}{F(D_c - H)} (\gamma_{\max} - \sin \gamma_{\max}) \tag{6.50}$$

where $\gamma_{\max}$ is calculated by Eq. (6.49).

During final molding of the profile the binder is redistributed within the article structure upon squeezing from the flanges and completes impregnation of the winding layer. The pressure generated during this process is calculated by Eq. (6.39). The axial constituent of the pressure forms an additional resistance force to pulling

$$P_{A0} = (p_r F_r + p_f F_f) \cdot (1 - \varphi_1) \tag{6.51}$$

where p_f and F_f are the pressure value and profile flange area, respectively.

Pressure values in the post and flanges differ because permeability in the longitudinal direction and in the binder bulk differ, too.

Calculations of adhesive forces effecting the article being formed should take into account that the material moving along the die heated to the binder curing point is subjected to polymerization. This process leads to the formation of a cellular structure of the polymer and its transition from a viscous liquid state to a solid state.

Analysis of chemorheological diagrams of the epoxy binder state[430] has shown (Figs. 3.1 and 3.2) that it transforms from the viscous-flow state to a highly elastic one during curing in the die. The adhesive tangential stresses in all sections of the die can be described by Eq. (5.27). The viscosity of the binder increases with increasing degree of polymerization and reaches its highest values as the adhesive stresses approach the binder shear threshold.

Experimental verification of the process has proved that a certain portion of the cured surface layer of the binder may breakdown during pultrusion. As a result, fine particles or grains of ruptured binder can be seen on the surface of the article leaving the die.

During curing the binder undergoes chemical shrinkage. This leads to reduction of the article in cross-section, and its deformation can be calculated by the formula

$$\varepsilon_{\mathrm{r}} = \varsigma\,(1 - \varphi_1) \tag{6.52}$$

where ς is the shrinkage factor of the binder.

The article undergoes chemical shrinkage gradually in the course of polymerization. This process goes hand-in-hand with relaxation of all initial stresses, and the ready article leaves the die at a 60 percent level of polymerization.[310,430] This means that there is no need for further molding, and the article can be polymerized fully in any heating device.

Consequently, the die length can be chosen with the presumption that adhesive stresses operate at its full stretch, i.e. the adhesive contact area equals to the total inner surface of the forming die. Assuming a linear variation in viscosity along the length of the die, we can substitute the adhesive tangential stresses with their mean value. The adhesive resistance force of the die can then be found from the relation

$$P_{\mathrm{af}} = 0.5\Pi L_{\mathrm{d}} C_1 \left(\eta + \eta_0^T\right) \tag{6.53}$$

where Π is the profile perimeter, and η_0^T is the binder viscosity in a viscoelastic state.

The total resistance to motion of the article being molded in all process stages is found as a sum of all constituents of the resistance and is to be equal to the force created by the pulling mechanism.

$$P_{ci} = \sum P_i \tag{6.54}$$

6.5 Conclusions

1. Based on the analysis of structural peculiarities of profile articles, the major stages of their force treatment during molding have been defined.

2. The dominant physical processes occurring at different stages of force treatment of composite complex-profile products have been studied. Binder squeezing in the interfiber gaps was found to play a determining role in force actions.
3. The main theoretical dependencies interrelating structural parameters of the molded material and force characteristics of the braiding and pultrusion processes have been derived.
4. The principal constituents of resistance forces to motion of the molded material have been examined at all stages of the technological process. Their relationship with force parameters of the process and design characteristics of the equipment used has been established.
5. Based on the analysis and experimental verification of the results obtained, a calculation procedure has been elaborated for the force parameters of the braiding–pultrusion process. Recommendations are presented for designing the technological rigging and equipment.

Chapter

7

Stitching of High-Modular Composite Structures at Molding Complex-Profile Ware

7.1 Design and Process Analysis: Choice of the Stitching Method

Profile walls in the junction with flanges can be stiffened by different methods and fixing elements. Stitching of a semifinished product by fine-fibrous threads or braids is believed to be advantageous to other methods as it lowers damageability of the work material during the technological process. Besides, stitching materials diminish stress concentration and fiber distortion when additional reinforcement has to be introduced and make homogeneous the adhesive bonds between the stitching, main fibers, and the binder.[354]

The problems of stitching by fibers or threads have received a great deal of attention in scientific literature.[191,200,237,388,401] A generalized analysis of investigations in this domain has led us to the following conclusions:

1. The principal mechanisms of interrelations between the material and tools have been comprehensively studied by the methods of the mechanics of deformable solids.
2. The stitching tools, including the needle, have been optimized and perfected as dependent on operation conditions and their designation.

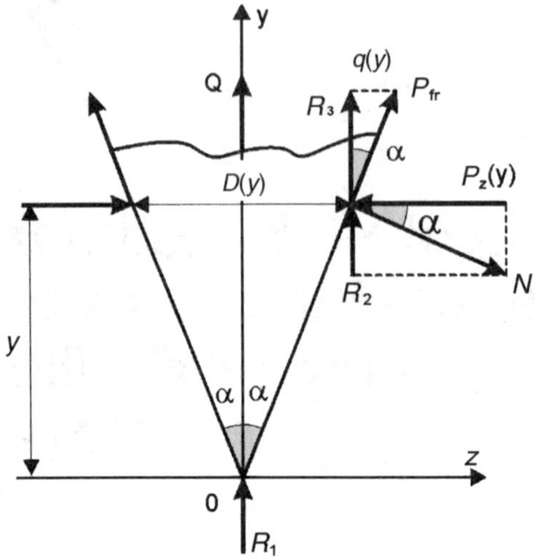

Figure 7.1 Stresses arising when the material is pierced by a needle.

3. The questions of a stable stitching process, damageability of the work material, quality of the stitch, etc. have been studied at length.
4. A series of standard piercing mills for different operation regimes has been worked out.

Stitching of semifinished composite ware is connected with specific features of the material, production process and physicomechanical properties of its ingredients.

Let us consider one of the design methods most frequently used for calculation of the stresses exerted on the needle (Fig. 7.1).[200,237]

During stitching, the needle moves in the direction of force Q and experiences the following resistance forces of the material: R_1 is the cutting force by the wedge blade; N is the force counteracting separating of the material; and P_{fr} is the friction force of the needle against the material.

The needle blade is not sharpened ideally but is somewhat rounded off, so the resistance R_1 appears at its penetration into the material. While overcoming this resistance, the needle ruptures fibers of the material.

The separation resistance N of material fibers is directed along the normal to the wedge inclines. It is a constituent of the horizontal separation resistance, which is proportional to penetration depth of the needle point into the material and can be determined by the relation

$$N = \frac{Ph}{\cos \alpha} \tag{7.1}$$

where P is the specific cutting resistance per unit depth of tip penetration; h is the tip penetration depth; and α is the needle tip angle.

The friction resistance P_{fr} is created by material pressure N on the wedge incline

$$P_{fr} = N \cdot f_{fr} \tag{7.2}$$

where f_{fr} is the friction coefficient of the tip over the material.

If to project all resistance forces in the direction of the motion of the needle tip, we must obtain the force Q needed for the tip to advance in the material bulk. Projections of named forces are calculated by the formulas

$$R_2 = N \sin\alpha = \frac{Ph \sin\alpha}{\cos\alpha} = Ph \tan\alpha \tag{7.3}$$

$$R_3 = P_{fr} \cos\alpha = f_{fr} N \cos\alpha = \frac{f_{fr} Ph \cos\alpha}{\cos\alpha} = f_{fr} Ph \tag{7.4}$$

Hence, the total resistance force will be calculated by the following equation:

$$Q = R_1 + 2R_2 + 2R_3 = R_1 + 2Ph(\tan\alpha + f_{fr}) \tag{7.5}$$

As it can be seen, the force needed to pierce the material with *a priori* known stiffness and thickness is largely dependent upon the point angle and shape as well as the friction coefficient of the tip against the material and needle size.

The formulas above work well for solid materials with low physicomechanical characteristics or fabric structures with knitting cells commensurable to or below the cross-size of the stitching tool. The threads that form the fabric commonly consist of short, strongly twisted fibers and can be therefore considered as an unbroken material. Under these conditions, an assumption of proportionality with respect to the separation resistance to the penetration depth of the needle may yield satisfactory results.

Reinforcing fillers of composite materials differ much from those considered above due, first of all, to their high mechanical characteristics, including strength and stiffness. To attain high mechanical properties of manufactured products, the filler fibers used should be kept straight.[150,222,348,406] This is why reinforcement threads for composite materials are either slightly twisted or straight at all and bonded through dressing or finishing agents. Hence, fibers are to be treated at stitching as isolated elements. Very often, a part or all fibers are impregnated by a binder before stitching, so friction forces are substituted for adhesive ones operating between the tool and binder.

The totality of the factors mentioned above necessitates additional analysis of Eq. (7.5) along with examination of the dependencies of the forces arising at needle penetration into materials of different structures.

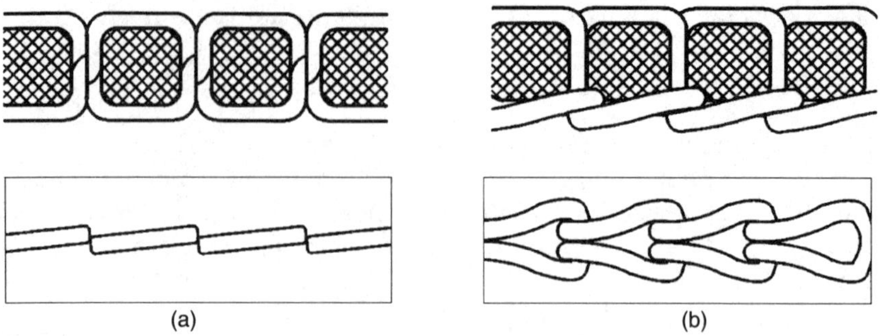

Figure 7.2 Types of stitches used: (*a*) double-thread stitch; (*b*) single-thread stitch.

Another critical feature of stitching is the condition of stitch interlocking. Great varieties of simple and figured stitches make use of one, two, and more threads (Fig. 7.2).

Most applicable is likely to be the double-thread scheme of stitching, during which the two threads are mechanically interwoven. The needle thread is fed to the interweaving zone when the material is pierced with a needle, and the shuttle thread is wound onto a bobbin found inside a traveling shuttle. When traveling back, the needle thread forms a loop into which the shuttle is passed, letting in the second thread. The loop slides over a smooth shuttle surface and, as they interweave, the needle thread draws the shuttle thread inside the material. When the interweaving is formed by two intersecting threads, the strength of the stitch is conditioned by that of the threads at cross-shear. To ensure this high level of consistency, the equipment used should be highly accurate and with a top surface finish.[417]

The thread in a single-thread scheme is drawn into the loop formed at piercing, which is supported and forwarded into the weaving zone by a special spring attachment. As a result, either a buttonhole or chain stitch is obtained. The strength of this stitch through the article thickness is much higher than that produced by a double-thread scheme and is twice as strong as that of the thread at tension.

The threads are first stretched in both schemes to tighten the stitch and consolidate the material being stitched.

Because the composite semi is fully or partially impregnated with a binder, both the needle and the needle's thread become soiled. The situation can be improved in the case where only the inner layers of the semi are impregnated and the post is not fully shaped, so that when the needle and thread pass through the external layers they are partly cleaned as a result of friction against nonimpregnated outer fibers. As for the double-thread scheme, it cannot be used for impregnated reinforced materials.

The single-thread scheme is considerably simpler, does not require high surface purity, and guarantees a high-strength stitch.

Testing of the technique has shown that, in some cases, the process of mechanically interlocking the seam turns out to be unnecessary. After the stitching process, the profile walls are completely shaped, the material is consolidated, and the winding layer is additionally impregnated. The loop formed during reciprocation of the needle is laid during further molding in the die on the article surface and impregnated with the binder. Under the action of friction and adhesive forces, the loop is shifted a little and removed from the piercing zone. It is also elongated and flattened, while the fibers of the stitching material line up and spread out over a considerable area, the material stitched becomes more compacted. It turns out after polymerization that the stitch strength is reached not only by mechanical engagement but also by strong adhesive bonding of the thread or braid with the article. This bond strength depends upon the maximum tangential stresses in the splice.

The third peculiarity of the stitching process of complex-profile rod articles produced by the braiding–pultrusion method is the condition of material motion.

In all stitching procedures, the material motion is precisely coordinated with that of the needle, being realized in the time interval when the needle is in its upper position. As soon as the material has advanced to the length of a stitch, the presser foot is let down, fixing the material in an immovable position, and the needle is then set in the material, which stays immovable for the time until interweaving is formed and the needle leaves it.

Pultrusion conditions require continuous motion of the material at all process stages. Thus stitching will bring about certain difficulties. In principle, there exist two approaches to this problem:

1. Elaboration of special equipment designs in which the article molded moves in coordination with the stitching unit in the moment the needle is in the material.
2. Development of special regimes of needle motion.

Hence, additional studies aimed at determining the stitch parameters and the stitching process as a whole are needed.

7.2 Analysis of Stitch Joint Strength

Because the realized stitch joint strength without the mechanical interlocking is conditioned by—first of all—magnitude and distribution of tangential stresses in the splice zone, the methods used to design adhesive junctions can be used in the analysis. As shown in Ref. 45, tangential stresses in the joint of two parts with constant length are found by the formula

$$\tau = Pk_2 \left[\frac{1 - \psi(1 - \cosh k_2 L)}{\sinh k_2 L} \cosh k_2 x - \psi \sinh k_2 x \right] \tag{7.6}$$

where

$$k_2 = \sqrt{\frac{G_s}{\delta_s} \cdot \left(\frac{1}{E_1 \delta_1} + \frac{1}{E_2 \delta_2} \right)} \qquad \psi = \frac{E_2 \delta_2}{E_1 \delta_1 + E_2 \delta_2} \tag{7.7}$$

where P is the load per unit width of the adhesive joint (in N/m); L is the joint length; E_1, E_2, δ_1, and δ_2 are the elasticity and thickness moduli of joined parts; and G_s and δ_s are the shear modulus and joint thickness, respectively.

It follows from Eq. (7.6) that the highest stresses acceptable for substitution in the strength condition are at the loop edges. Under the known condition of $E_1\delta_1 < E_2\delta_2$, the maximum stress values are calculated by the equation

$$\tau_{max} = Pk_2 \frac{\psi + (1 - \psi) \cosh k_2 L_{lim}}{\sinh k_2 L_{lim}} \leq [\tau_s] \tag{7.8}$$

where $[\tau_s]$ is the binder shear strength.

To avoid twitching of the threads or braids as a result of rupture of the adhesive layer, it is desirable to obey the full-strength condition of the adhesive joint and the breaking strength of the stitching material. So instead of P, its maximum possible value should be substituted into the strength condition Eq. (7.8)

$$P_{max} = \sigma_w^0 \frac{F_w}{b_w} = \frac{2\, T\sigma_w^0}{\gamma_w b_w} \cdot 10^{-6} \tag{7.9}$$

where b_w and F_w are the width and area of the thread; σ_w^0 is the thread tensile strength; and T_w and γ_w are the linear (tex) and specific (kg/m^3) densities of the thread.

The total area of the fibers of threads or braids F_t is calculated by the formula

$$F_w = \frac{T_w}{\gamma_w} \cdot 10^{-6} (\text{m}^2)$$

Thus the minimal loop length that obeys the full-strength condition is found from the inequality

$$\frac{\psi + (1 - \psi) \cosh k_2 L_{min}}{\sinh k_2 L_{min}} \leq \frac{\gamma_w b_w [\tau_c]}{2k_2 T_w \sigma_w^0} \cdot 10^6 \tag{7.10}$$

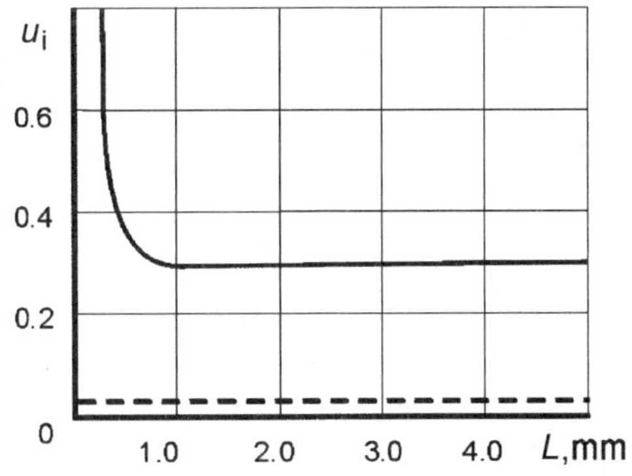

Figure 7.3 To the calculation of the full-strength condition of a seam joint: (—) left-hand side of the inequality; (- - -) right-hand side of the inequality.

The left- and right-hand sides of the inequality obtained are presented in Fig. 7.3 as a function of the loop length for stitching carbon plastics by a carbon braid. It follows from the diagrams that these functions have no points of intersection, i.e., the given inequality has no solution. Analogous results have been derived for other types of materials.

It is evident that the full-strength condition is improbable for the stitching method chosen. With loop lengthening above some value, the maximum tangential stresses drop to a certain limit and further elongation of the loop does not in fact influence their magnitude.

To estimate the bearing capacity of the junctions of similar materials, the limiting length $L_{\lim}$ is usually taken so as to minimize the tangential stresses in the middle of the junction to below 1 percent of their maximum value.

To determine the location of the minimum function of tangential stresses for unlike materials, one should take its first derivative and equate it to zero. Solution of the relation obtained relative to the coordinate of the minimum x_0 gives

$$x_0 = \frac{1}{k_2} \tanh \left[\frac{\psi \sinh k_2 L}{1 - \psi(1 - \cosh k_2 L)} \right] \tag{7.11}$$

Analysis of the relation (7.11) shows that the coordinate of the minimum of tangential stresses under specified conditions is also a function of the adhesive joint length. A diagram of x_0 versus L dependence is shown in Fig. 7.4.

The diagram proves that, in contrast to the adhesive joint with similar geometrical parameters and stiffness, the minimum is not found in the joint middle but is shifted and that the shift value from the middle position depends on the length parameter.

When this function is not absolute but a relative extreme coordinate x_0/L, the dependence is presented more clearly. The diagram of such dependence is shown in Fig. 7.5.

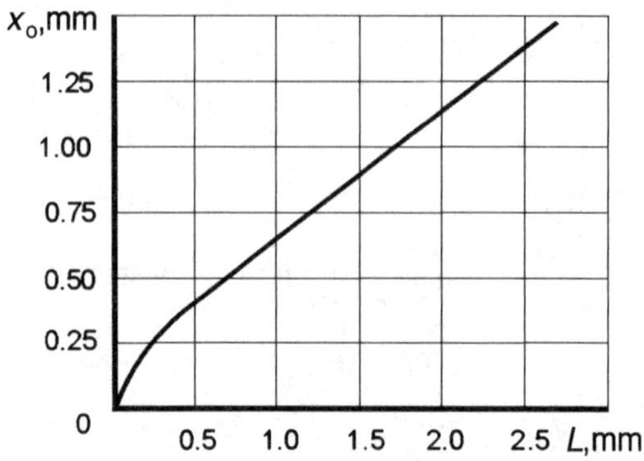

Figure 7.4 Effect of the loop length on coordinates of the minimum function of tangential stresses.

The function of the extremum has been defined over the whole interval considered. Under certain length values, the tangential stresses tend to zero, while the function arth $k_2x \to 1$ and is consequently undetermined. To calculate limiting values of the efficient joint length, one should evidently choose a ratio of x_0/L that is close to the determination bounds of this function.

Based on the analysis performed, a conclusion can be made that, in case characteristics and dimensions of adhesive joints differ, it is possible to use $x_0 = 0.5L_{\text{lim}}$ as a coordinate of the minimum.

The relation of the minimum of tangential stresses to their maximum can be found by substitution of the above value in Eq. (7.6); after certain manipulations, we have

$$\nu(L) = \frac{\tau(x = 0.5L)}{\tau_{\max}(L)} = \frac{\cosh \frac{k_2L}{2}}{\psi + (1 - \psi)\cosh k_2L} \tag{7.12}$$

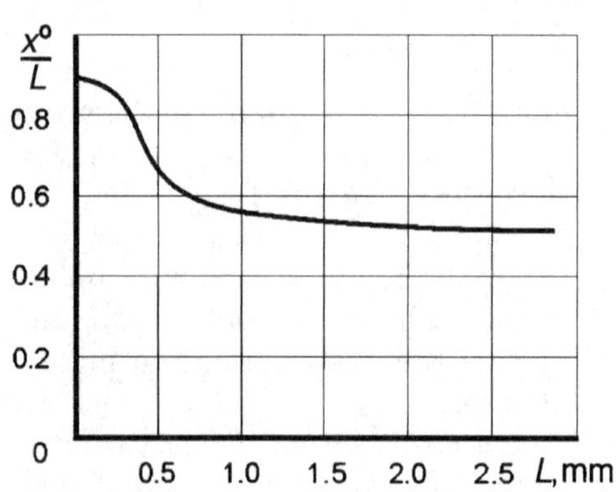

Figure 7.5 Location of the relative minimum of the function of tangential stresses versus loop length.

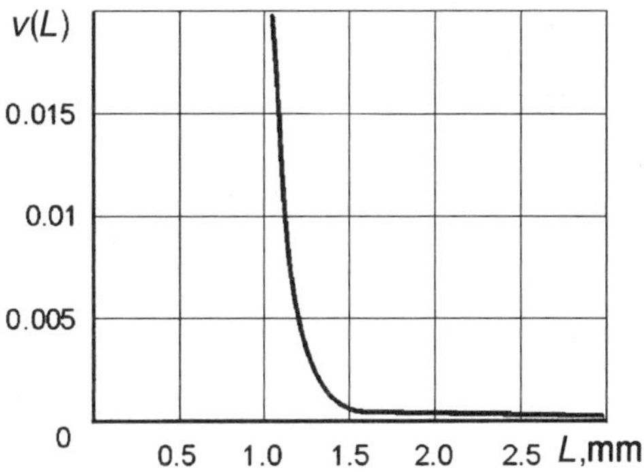

Figure 7.6 Relative minimum of tangential stresses as a function of stitch loop length.

The dependence of relative minimum of tangential stresses versus loop length of the stitched joint calculated by Eq. (7.12) is illustrated in Fig. 7.6. It follows that the minimal stresses reach 1 percent of the maximum value at a splice length 1.2 mm; when the splice is 2 mm long, they approach zero. It is evident that longer loops are inexpedient because they do not strengthen the joint.

If to substitute the L_{lim} value obtained in Eq. (7.8), we can obtain the minimum possible maximum of tangential stresses. Based on Eq. (7.9), where strength σ_{w}^{0} should be substituted by $[\sigma_{\text{w}}]$, we can calculate the maximum bearing capacity of the joint

$$[\sigma_{\text{w}}] = \frac{[\tau_{\text{c}}]\gamma_{\text{w}} b_{\text{w}} \sinh k_2 L_{\text{lim}} \cdot 10^6}{2T_{\text{w}} k_2 [\psi + (1-\psi)\cosh k_2 L_{\text{lim}}]} \tag{7.13}$$

$[\sigma_{\text{w}}]$ is the strength value that should be substituted in Eq. (7.40) to calculate the joint parameters for the accepted stitching method.

In case the stitch is bunched mechanically, the strength of the joint will depend on that of the sewing material at cross-shear

$$[\sigma_{\text{w}}] = [\tau_{\text{w}}] \tag{7.14}$$

7.3 Kinematic and Force Analysis of Stitching Braided Structures (Based on High-Modular Reinforcing Fibers)

When an unsaturated composite material undergoes stitching, both the material and the tool are subjected to reactive forces generated between the components, along with elastic and friction forces.

Elasticity moduli of the fibers of braided composite structures are by far higher in contrast to those appearing in textile threads. This is why the fibers can be considered in calculations to be nonstretched ones so that only their flexure parameters will be considered.[348]

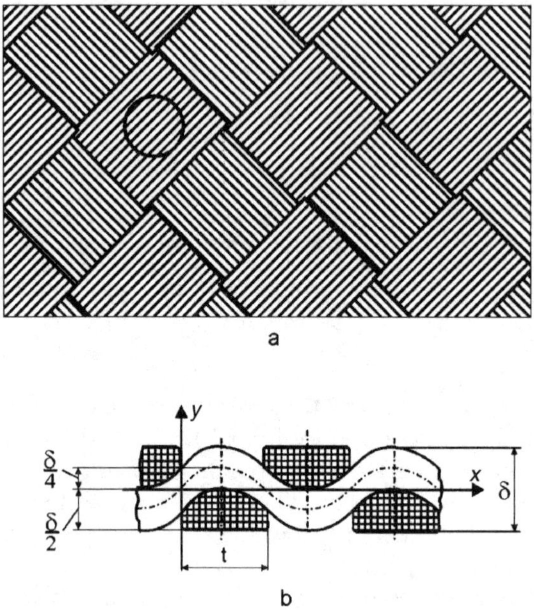

Figure 7.7 Braided structure model of a composite material: (*a*) general view; (*b*) lengthwise section of a fiber.

During formation of the braiding layer, the braids form an interwoven structure like that presented in Fig. 7.7.

It is seen (Fig. 7.7(*b*)) that each fiber is deformed at braiding and takes a sinusoidal shape of period $2t$ and amplitude $\delta/4$, where t corresponds to the reinforcing braid width and δ is the braided layer thickness.

The flexure function y of the fibers will be

$$y = f_0 \sin \frac{\pi x}{t} \tag{7.15}$$

When displacements are insignificant, the sinusoid can be substituted by a parabola. In this case, the maximal flexure of the fibers can be determined by the relation

$$y_{\max} = f_0 = \frac{\delta}{4} = \frac{P_0 t^3}{48 E I_f} \tag{7.16}$$

where $I_f = \pi d^4/64$ is the inertia moment of the fiber cross-section and d_f is the fiber diameter.

The force applied onto the fiber to attain a given flexure value is

$$P_0 = \frac{3\pi E \delta d_f^4}{16 t^3} \tag{7.17}$$

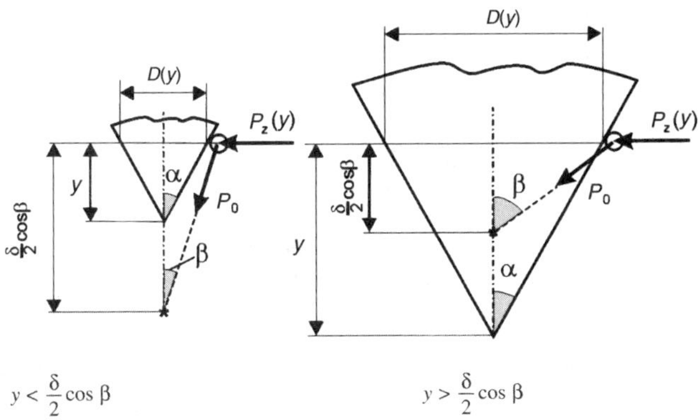

Figure 7.8 Changes in the fiber flexure plane at different stages of needle penetration into the material structure based on high-modular fibers.

The needle inserted into the material structure changes the flexure plane of the fibers. This plane turns to some angle β whose value depends on the needle size. The pressure exerted by the fibers varies and depends on the needle penetration depth y (Fig. 7.8):

$$P_z = P_0 \sin\beta = \frac{4P_0 \tan\alpha}{\delta} y \tag{7.18}$$

where α is the needle angle.

When the diameter of the needle is $D \to \delta/2$ and the angle $\beta \to \pi/2$, then $P_z \to P_0$. As the needle penetrates deeper and its diameter enlarges, further deformation of the fiber occurs due to either partial or full straightening of their flexure sinusoid, which is commonly accompanied by slippage of these fibers relative to transverse ones. As a result, an additional constituent of the transversal resistance is formed.

When longitudinal forces are active in the fibers, the maximal flexure is interrelated with that of the nonstretched fibers through the relation

$$f = \frac{f_0}{1 + \frac{S}{P_S}} \tag{7.19}$$

where S is the longitudinal force and $P_S = \pi^2 EI_f(\mu t)^2 = \pi^3 E d_f^4/64t^2$ is Euler's force.

Longitudinal forces alleviate flexure of the fibers by a value $\Delta y = f_0 - f$, but at the same time, they increase the flexure of transversal braids by a similar value (Fig. 7.7(*b*)):

$$f^* = f_0 + (f_0 - f) = 2f_0 - f = f_0 \left(2 - \frac{1}{1 + S/P_S}\right) \tag{7.20}$$

Longitudinal forces also appear in the fibers of transversal braids owing to raised resistance to the deflection line length. Hence, the cross-force affecting nonstretched fibers generates the flexure

$$f_0^* = f^* \left(1 + \frac{S^*}{P_S}\right) \tag{7.21}$$

where S^* is the longitudinal force operating in the transverse fibers.

This force arises due to friction forces in the fibers and is given by

$$S^* = F_{fr} = f_{fr} P_0^* = \frac{48 f_{fr} E I_f f_0^*}{t^3} \tag{7.22}$$

where f_{fr} is the friction coefficient.

If we solve the equation in relation to S^* and substitute Eq. (7.21) in Eq. (7.22), we obtain

$$S^* = \frac{P_S}{\frac{\pi^2 t}{48 f_{fr} f^*} - 1} \tag{7.23}$$

In this equation, f^* is found from Eq. (7.20)

$$\text{as } S \to 0, \quad f^* \to \delta/4, \quad \text{and as} \quad S \to \infty, \quad f^* \to \delta/2$$

Variation in the rate of S^* from the minimum to the maximum depends on structural parameters and friction coefficient. The calculation error of the efficient length of deformed fibers with different f^* does not surpass as a rule 5 percent. So far, flexure of transverse fibers can be taken in engineering estimates of the friction forces as maximal and equal to

$$f_{\max}^* = 2 f_0 = \frac{\delta}{2}$$

The maximal longitudinal force in fibers of the transverse braid is then

$$S_{\max}^* = F_{fr}^{\max} = \frac{P_S}{\frac{\pi^2 t}{24 f_{fr} \delta} - 1} \tag{7.24}$$

Under stronger longitudinal force S^*, the fibers start to slip and their flexure amplitude increases. The same occurs with the longitudinal fibers, although their flexure amplitude diminishes as the longitudinal force grows until the braid becomes straight (Fig. 7.9).

The friction force between fibers is a constant parameter under the assumptions accepted. This force induces the variable longitudinal force in the threaded fibers (Fig. 7.10).

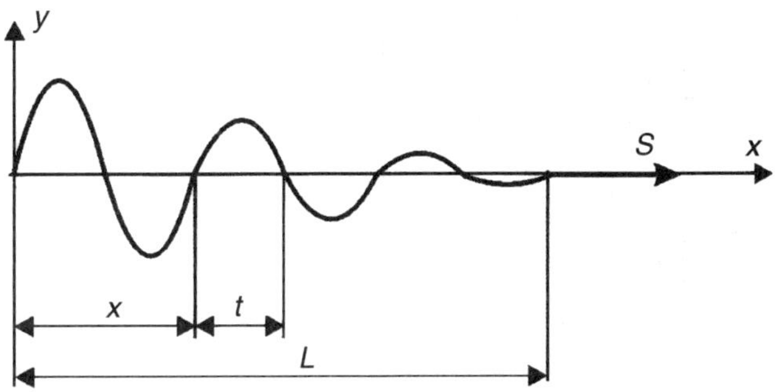

Figure 7.9 Variations of the braid amplitude at slippage under longitudinal force action.

Its magnitude can be found in terms of the equilibrium condition

$$dS = F_{\text{fr}}\frac{dx}{t}$$

Then,

$$S = \int \frac{F_{\text{fr}}}{t}dx = \frac{F_{\text{fr}}}{t}x + C$$

The integration constant is obtained from the initial condition at $x = 0$, $S = F_{\text{fr}} = C$.

$$S = \frac{F_{\text{fr}}}{t}x + F_{\text{fr}} = F_{\text{fr}}\left(\frac{x}{t} + 1\right) \tag{7.25}$$

The maximum flexure of the fiber stretched by force S to a distance x is

$$f = \frac{f_0}{1 + \frac{F_{\text{fr}}}{P_{\text{S}}}\left(\frac{x}{t} + 1\right)} \tag{7.26}$$

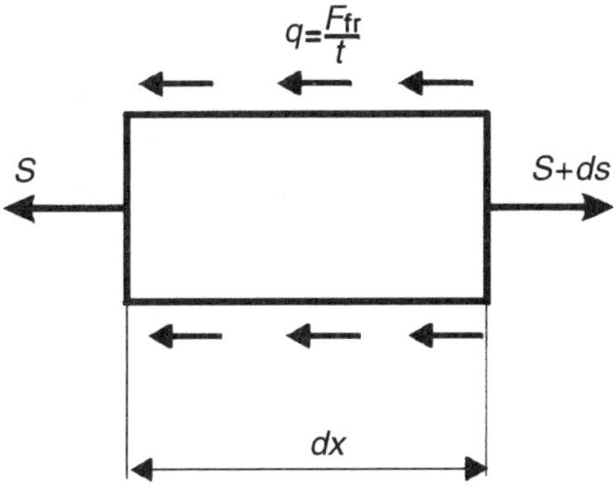

Figure 7.10 Design model of the resistance force of a braid at slippage.

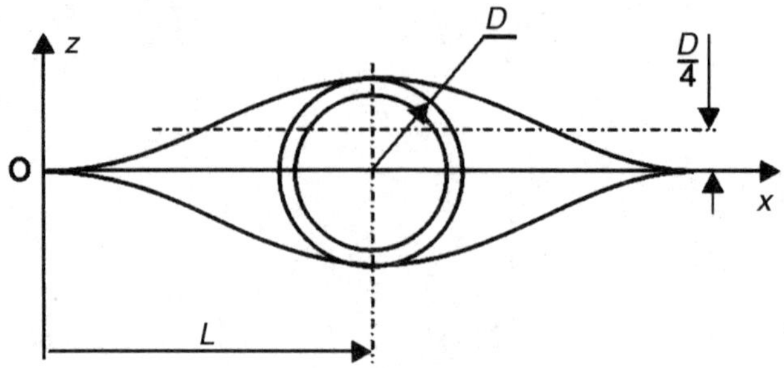

Figure 7.11 Deformation diagram of fibers at penetration of the needle.

The total increment of fiber length through braiding will be

$$\Delta = \frac{1}{2}\int_{0}^{L} (y')^2 dx \tag{7.27}$$

Presuming that the flexure function y of the nonstretched braid fibers is described by Eq. (7.15), this function for the stretched fibers of the braid will be

$$y_{\mathrm{d}} = f \sin \frac{\pi x}{t} \tag{7.28}$$

The total increment of the braid length with varied function is

$$\Delta - \Delta_{\mathrm{d}} = \frac{1}{2}\int_{0}^{L} (y')^2 dx - \frac{1}{2}\int_{0}^{L} \left(y'_{\mathrm{d}}\right)^2 dx \tag{7.29}$$

This increment in the length of the braid should be similar to that of the deflection curve in the z axis at penetration of a needle of diameter D (Fig. 7.11).

If the flexure curve function along the z axis varies by a sinusoidal law,

$$z = \frac{D}{4} \cos \frac{\pi x}{L} \tag{7.30}$$

the increment of the curve length can also be found by the formula

$$\Delta_z = \frac{1}{2}\int_{0}^{L} (z')^2 dx \tag{7.31}$$

If we equate the right-hand sides of Eqs. (7.29) and (7.31) and then shorten and differentiate the relation obtained with respect to x, we obtain the dependence

$$(y')^2 - (y'_d)^2 = (z')^2 \tag{7.32}$$

Having found derivatives and squares of the functions, we then have

$$\frac{\pi^2 f_0^2}{t^2}\cos^2\frac{\pi x}{t} - \frac{f_0^2 F_{fr}^2 \sin^2\frac{\pi x}{t}}{P_S^2 t^2\left[1+\frac{F_{fr}}{P_S}\left(\frac{x}{t}+1\right)\right]^4} - \frac{\pi f_0^2 F_{fr}\sin\frac{2\pi x}{t}}{P_S t^2\left[1+\frac{F_{fr}}{P_S}\left(\frac{x}{t}+1\right)\right]^3}$$

$$- \frac{\pi^2 f_0^2 \cos^2\frac{\pi x}{t}}{t^2\left[1+\frac{F_{fr}}{P_S}\left(\frac{x}{t}+1\right)\right]^2} = \frac{\pi^2 D^2}{16L^2}\sin^2\frac{\pi x}{L} \tag{7.33}$$

Equation (7.33) will be true under all x values, i.e., over the entire length of the deformed braid. Let us take $x = t$. Then, all terms containing the function $\sin\frac{\pi x}{t}$ will be equal to zero and $\cos\frac{\pi x}{t} = 1$.

If to express L through the influence function equal to the number of braiding steps participating in deformation $L = kt$, then after substitutions Eq. (7.33) will take the form

$$\frac{\pi^2 f_0^2}{t^2} - \frac{\pi^2 f_0^2}{t^2\left[1+\frac{2F_{fr}}{P_S}\right]^2} = \frac{\pi^2 D^2}{16k^2t^2}\sin^2\frac{\pi}{k}$$

After shortening, and bearing in mind that $f_0 = \delta/4$, we finally arrive at

$$\frac{1}{k}\sin\frac{\pi}{k} = \frac{\delta}{D}\sqrt{1-\frac{1}{\left(1+\frac{2F_{fr}}{P_S}\right)^2}} \tag{7.34}$$

Further, $F_{fr} = F_{fr}^{max}$ determined by Eq. (7.24) is substituted into Eq. (7.34). Although Eq. (7.34) has no analytical solution, it can be solved by numeric or graphic calculations.

Analysis of the diagram of the left-hand side of this equation, i.e., function $u = \frac{1}{k}\sin\frac{\pi}{k}$ (Fig. 7.12) shows that Eq. (7.33) can be applied only when $k \geq 1.5$.

By solving Eq. (7.34) with respect to the needle diameter D, we come to a relation

$$D = \frac{k\delta}{\sin\frac{\pi}{k}}\sqrt{1-\frac{1}{\left(1+\frac{2F_{fr}}{P_S}\right)^2}} \tag{7.35}$$

The diagram of the influence parameter k dependence on the needle diameter D is shown in Fig. 7.13. It is seen that diameters of the needle corresponding to the region of k practical values are found within $D \geq 0.5\delta$.

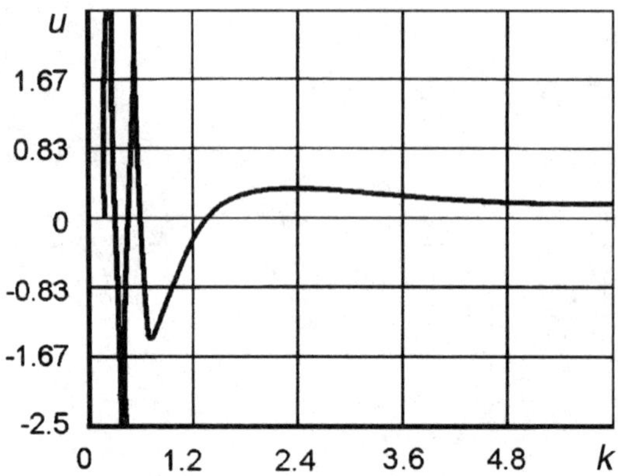

Figure 7.12 Diagram of the left-hand side function of formula (7.34).

Based on known k for given parameters of the braiding material (f_{fr}, δ, t) and needle diameter D, one can find the efficient length L and, consequently, fiber flexure over the z axis

$$z_{max} = \frac{P_z^* L^3}{48EI_f} = \frac{D}{4}\left[1 + \frac{F_{fr}}{P_S}\left(\frac{L}{t} + 1\right)\right] = \frac{D}{4}\left[1 + \frac{F_{fr}}{P_S}(k+1)\right] \quad (7.36)$$

The motive, transversal force of the resistance of deformed fibers and the friction force affect the needle inserted in the braided layer (Fig. 7.1). Because a sharpened needle has a variable cross-sectional diameter lengthwise $D(y)$, the pressure against the needle $P_z^*(y)$ varies in response to location of the considered fiber layer. The amount of fibers in the layer can be determined as $n = t/d_f$, providing that the pressure is effective from both ends of the needle and the thickness of the layer is d_f.

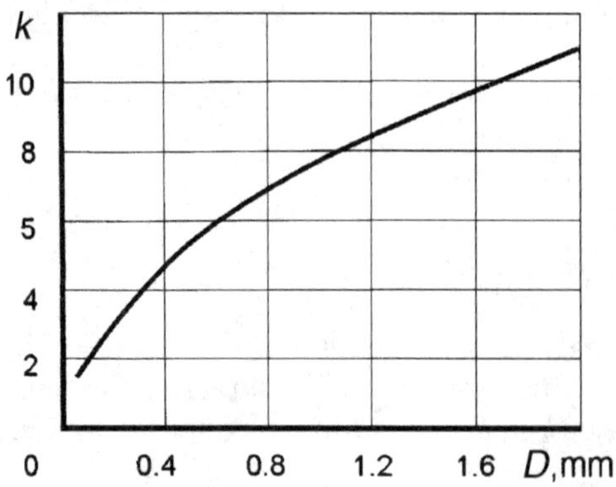

Figure 7.13 Variations of the influence parameter versus needle diameter.

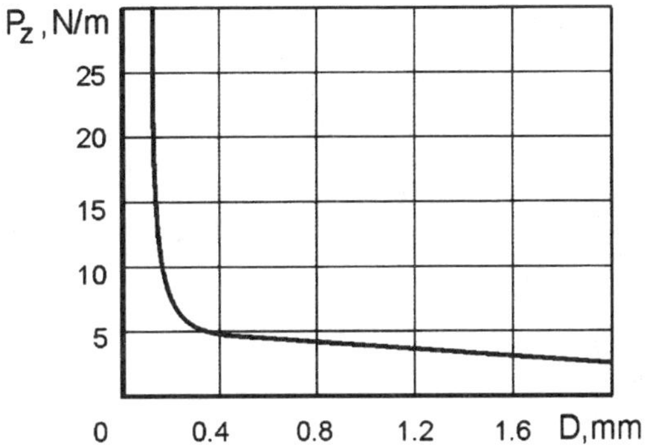

Figure 7.14 Linear force of transversal pressure of the fibers versus current diameter of the needle.

Hence, the additional force per unit length induced by slippage resistance of the fibers within the material structure can be calculated by the equation

$$P_z^*(y) = \frac{3\pi\, ED(y)\, d_{\mathrm{f}}^2}{16\, k^3(y)\, t^2}\left[1 + \frac{k(y)+1}{(\pi^2 t/24\,\delta\, f_{\mathrm{fr}}) - 1}\right] \tag{7.37}$$

As it follows from the analysis of Eqs. (7.35) and (7.37), values of both $D(y)$ and $P_z^*(y)$ are the functions of the influence parameter k. Therefore, the dependence of $P_z^*(y)$ on $D(y)$ can be presented in parametric form.

This dependence is illustrated graphically in Fig. 7.14. It is evident that the force exerted by the needle on the fibers grows with its diameter.

At $k \le 1.5$, corresponding to diameters $D \le 0.5\delta$, the formulas above are not true because the mechanism of reaching respective flexure changes. As indicated earlier, although slippage of the fibers does not occur at this diameter, the flexure plane of the layer into which the needle penetrates does change. It turns by an angle β that depends on sharpening of the needle (Eq. (7.18)) and its penetration depth (Eq. (7.8)) into the material structure. Because the needle scarcely ever acquires an ideally sharpened tip whose radius enlarges during operation as it gets blunted, the actual angle α_{R} varies from zero to a needed value. Thus, some of the fibers are in the zone of angles less the critical one and rupture. To avoid or at least abate this phenomenon, a special sharpening method is used. It consists in sharpening of the needle at a larger angle as compared to α. This leads to less damage to fibers and the needle dulls more slowly.

The maximum normal pressure exerted on the fibers by the needle is dependent in this case only on the lip angle of the needle α^* and cannot exceed their shear strength. This pressure with account of the friction forces can be determined by the formula

$$N^{\max} = \frac{P_0}{\cos\alpha^* - f_{\mathrm{fr}}\sin\alpha^*} \le \frac{\tau_{\mathrm{b}}^0 \pi d_{\mathrm{f}}^2}{4} \tag{7.38}$$

where P_0 is found from Eq. (7.17).

Proceeding from Eq. (7.38), we can define limiting α^* value at which fibers will not rupture

$$\alpha^* \le \operatorname{acos}\left(\frac{\frac{4P_0}{\pi d_{\mathrm{f}}^2 \tau_{\mathrm{b}}^0} + f_{\mathrm{fr}}\sqrt{1 + f_{\mathrm{fr}}^2 - \left(\frac{4P_0}{\pi d_{\mathrm{f}}^2 \tau_{\mathrm{b}}^0}\right)^2}}{1 + f^2}\right) \tag{7.39}$$

For a braided structure of carbon fibers ($d_{\mathrm{f}} = 6 \cdot 10^{-3}$ mm, $E = 300{,}000$ MPa, $f_{\mathrm{fr}} = 0.1$, $\tau_{\mathrm{b}}^0 = 1250$ MPa), $\alpha^* = 84.3°$. As long as the needle lip angle is in fact much less, the fibers do not rupture even at $2\alpha^* \approx 90 - 120°$.

The transverse force acting on the needle during stitching is not stable over all zones, excluding rather large D values; consequently, the longitudinal forces can be found by Eq. (7.3) through Eq. (7.5).

In addition, axial constituents of the forces with known functions of transversal forces per unit length can be determined by integrating over the penetration depth of the needle

$$Q = \int_0^{\delta} P_z(y)\,(\tan\alpha + f_{\mathrm{fr}})dy \tag{7.40}$$

The linear force will be found by multiplication of the force value t/d_{f}^2 acting on a single fiber. Hence, the initial piercing force R_1 will be

$$R_1 = \int_0^{y_1^{\max}} \frac{2P_0\, t \cdot \tan\alpha}{\delta\, d_{\mathrm{f}}^2}(\tan\alpha + f_{\mathrm{fr}})dy \tag{7.41}$$

At $y_1^{\max} = \frac{\delta}{2\tan\alpha}$, $R_1 = \frac{P_0 t \delta}{2d_{\mathrm{f}}^2}(1 + \frac{f_{\mathrm{fr}}}{\tan\alpha})$.

To find forces $(R_1 + R_2)$, the value $P_z(y)$ is calculated by summing up the forces arising from the flexure plane deviation of the fibers and their slippage:

$$P_z(y) = P_0 + P_z^*(y) \tag{7.42}$$

The integral function is easily found via integration with respect to the influence parameter k whose limits are conditioned by the maximum diameter of the needle at piercing through the entire packet of the material (Fig. 7.13).

The structure of the formulas obtained for impregnated composite materials remains the same except for the friction forces occurring between the fibers. They are substituted by those of adhesive bonds, which are a function of the binder viscosity. So, instead of $S = F_{\text{fr}}(\frac{x}{t} + 1)$, in all formulas we use the longitudinal force value acting on a fiber

$$S = \pi\, d_{\text{f}}\tau_{\text{at}} \cdot x \tag{7.43}$$

where τ_{at} is calculated by Eq. (5.27).

7.4 Calculation of Forces and Pressures Exerted on the Sewing Material at Stitching

The sewing material represents as a rule a slightly twisted fine-fiber thread, strand or a roving that can be easily ruptured under loading.

Mainly three types of forces affect fibers of the sewing material:

1. The pretension force of a thread or braid specified by strainers.
2. Inertia forces arising at accelerated motion of the threads.
3. Pressure and friction forces appearing between the needle and fibers of the material.

The pretension force in the absence of stitch bunching is usually negligible and is largely intended to eliminate sagging and entanglement of the sewing material during processing.

In case of mechanical interlocking of the stitch, this force is aimed at straightening the upper base and lower loop as well as compaction of the work material. The pretension force P_{PS} is a process parameter that can be regulated within the specified limits.

The dynamic constituent of the forces depends on kinematic parameters of stitching, including motion of the needle when piercing the material structure, velocity, and acceleration of a thread or braid displacement.

The known laws of motion used when designing various mechanisms[234] are connected, usually with the presence of some type of impact loads.

So-called rigid impact may take place when the material velocity varies up to a finite value in discontinuous leaps, i.e., within a short time interval. Acceleration in this case tends to infinity and reaches values large enough in practice to lead to fast refusal of mechanisms. This law of motion is rarely used in designing machinery.

The phenomenon of soft impact is attributed to jerky variation of acceleration by a finite value or in a short time interval through either linear, parabolic, cosine, or some other law of motion. This type of impact is considered admissible, and mentioned motion laws are frequently employed in different mechanisms.

One of most shock-free laws is the sinusoidal law of acceleration variation. This is precisely the law that is used for stitching devices designated for processing easily damaged fibrous materials. Impact loads are not completely excluded at stitching.

As the needle moves in its upper position, its eye slides against a thread remaining practically immovable in this moment. As the needle moves down, the thread or braid moves for some time together with it, and its velocity is almost equal to that of the needle. As soon as the needle pierces the material and passes the thread through it, the needle catches the upper thread from both advancing and retracting groove sides. Because of slack in the upper base of the stitch, excess thread appears near the retracting groove. Simultaneously, the lower loop is tightened round the needle found inside this loop. The velocity of the thread through the eye V_f grows smoothly and reaches a value twice as high as that of the needle V_N in the moment the needle fixes the thread of the upper stitch base (Fig. 7.15).

For the sinusoidal law of motion,

$$x = \frac{H}{2} \sin \omega t \tag{7.44}$$

Here, H is the full speed of the needle; $\omega = \pi n/30$ is the circular frequency; and n is the rotational velocity of the main shaft of the stitching device.

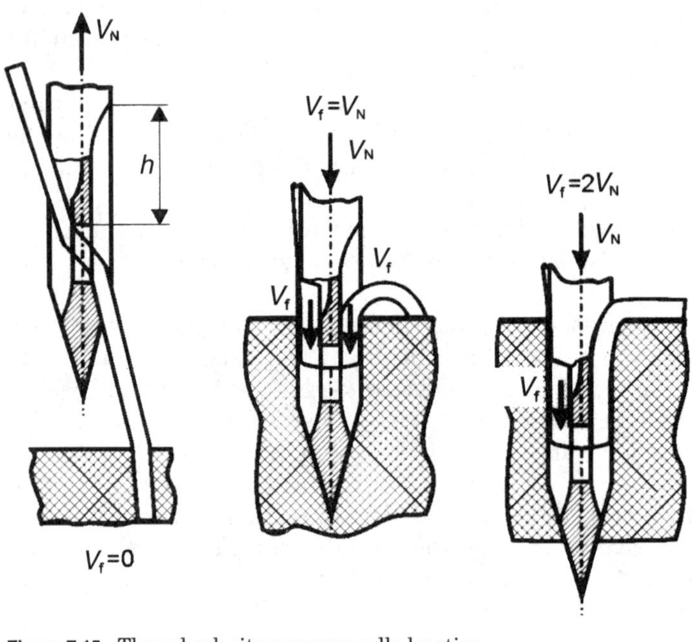

Figure 7.15 Thread velocity versus needle location.

To avoid rigid impact, a special recess is made in the needle whose size h guarantees the time needed for velocity growth and, consequently, acceleration.

$$t = \frac{1}{\omega}\text{asin}\,\frac{2h}{H} \tag{7.45}$$

It is difficult to define the actual law of velocity variation of the thread, so we presume in calculations that it keeps to the linear law, where acceleration is constant and is equal to

$$a_{\text{m}} = \frac{H\omega}{2t}\cos\omega t \tag{7.46}$$

The dynamic constituent of the thread tension force is found from the relation

$$P_{\text{df}} = m_{\text{w}}a_{\text{mp}} = \rho_{\text{w}}F_{\text{w}}L_{\text{fl}}a_{\text{m}} = \frac{TL_{\text{fl}}a_{\text{m}}}{g}\cdot 10^{-6} \tag{7.47}$$

where $F_{\text{w}} = \frac{T}{\gamma_{\text{W}}}\cdot 10^{-6}$ is calculated by formula (4.39); L_{fl} is the free length of the thread from the strainer to the needle eye; and g is the gravitational acceleration.

As the thread or braid bends round the eye edge, it experiences friction forces whose value can be determined by Euler's formula

$$P_{\text{fr}} = (P_{\text{PS}} + P_{\text{df}})\cdot e^{\alpha f} \tag{7.48}$$

where $\alpha = \pi$, which is the wrap angle of the eye edge by the thread.

The total force applied on the thread by the needle to make it move is then

$$Q_{\text{W}} = (P_{\text{PS}} + P_{\text{df}})\cdot(2 + e^{\pi f}) \tag{7.49}$$

This force is applied perpendicularly to the fibers, i.e., is transversal with respect to them, and can lead to breakage of the thread fibers as a result of shear. So, the strength condition should be written as

$$\tau_{\max} = \frac{Q\,\gamma_{\text{w}}\cdot 10^{6}}{T} \le \tau_{\text{W}}^{0} \tag{7.50}$$

Proceeding from this relation, the maximum allowable force of a thread pretension is

$$P_{\text{PS}} = \frac{T\,\tau_{\text{W}}^{0}\cdot 10^{-6}}{\gamma_{\text{w}}(2 + e^{\pi f})} - P_{\text{df}} \tag{7.51}$$

As far as force Q_{w} is applied on the needle by the thread, it should be added to the piercing forces as a constituent of the total resistance to needle motion.

7.5 Conclusions

1. The design analysis conducted herein has shown the most important features of the stitching process of composite materials, as a result of which a single-thread scheme using the buttonhole stitch model has been chosen.
2. Strength analysis of the adhesive joint has shown that it is admissible in some cases to use a simplified scheme of the joint without mechanical interlocking of the stitches. This makes it possible to employ simpler process equipment to provide strong enough joints.
3. An evaluation procedure has been developed to study the bearing capacity of the stitch joint without the mechanical interlocking that interrelates strength and design parameters of the joint.
4. The existing procedures of analyzing and designing the forces and pressures that affect stitching processes were found to be inapplicable at manufacture of composite products incorporating high-modular, fibrous, reinforcing fillers. A new design method that takes into account the deformation processes related to variations in braiding amplitude and slippage of the braiding structure elements has been worked out.
5. A calculation method of pretension forces of threads and braids that takes into account the dynamic and kinematic factors of the stitching process has been developed.

Chapter 8

Development of the Production Process for Complex-Profile Composite Wares

8.1 General Characteristics of the Process

Based on the analysis of structural peculiarities of the complex-profile rod elements formed by joined operations of braiding and pultrusion, the following sequence of stages of the production process has been worked out:

1. Preparatory stages consisting of
   - Preliminary sampling and input control of the main and auxiliary reinforcement materials
   - Preparation of binder components and a compound of a needed composition
   - Hold-down of required amount of bobbins with the main reinforcement on bobbin carriers, as well as bobbins with prepregs and braiding reinforcement on the braider
   - Preparation of the equipment and tooling for operation, adjustment of strainers, impregnation and squeezing devices, distributors, braiders and setting of temperature regimes in the forming die and thermal chamber
   - Threading of the reinforcing material and fixing in the pulling mechanism
2. Pretension of the main reinforcement of the flanges, its impregnation with the binder, and squeezing of binder excess.

3. Spreading of the flange and post reinforcement over the mandrel surface.
4. Molding of impregnated reinforcement on the mandrel in the preforming die and manufacture of a cylindrical semifinished product.
5. Laying of the braiding layer on the semi surface.
6. Calibration of the semi to reach the needed outer perimeter of the article.
7. Preforming of the article post.
8. Stitching of the profile article in the junction between the post and flanges.
9. Final molding of the profile post.
10. Final molding of the article profile in the die.
11. Partial curing of the complex-profile article in the die.
12. Final curing of the profile article in a thermal chamber.
13. Finishing operations
    - Machining, including cutting to a needed size
    - Visual quality control
    - Test sampling of the chief physicomechanical characteristics of the product
    - Marking and packing

This process includes in fact all main stages of pultrusion of unidirectional composites but for additional stages of forming a braided layer as well as preliminary molding and stitching of the semi. So far, the major requirements imposed on both processes and equipment are almost the same.

Nevertheless, structural peculiarities of complex-profile articles and process features already considered in Chap. 6 necessitate certain adjustments in the manufacturing stages. These are primarily the preforming parameters and calibration of a semi along with the molding regimes of the braided layer. Selection criteria of the process parameters of the braided layer differ from analogous molding regimes of tubular and solid products obtained by the braiding–pultrusion process, so they cannot be simply transferred but need some corrections.

In addition, a series of additional parameters concerning stitching of the semi at the stage of post preforming must be considered. They are kinematic properties, force parameters, and design solutions of the stitching device.

Initially, one should know the dimensions and structural parameters of the article being manufactured. Under known loading behavior, the structural and geometrical parameters of the complex-profile rodlike composite elements can be calculated by formulas (4.23) to (4.25), (4.28), (4.33), (4.35), and (4.40). Simultaneously, the type of reinforcing fillers (see Sec. 4.3) and the binder are chosen considering the general concepts of the binders used in pultrusion processes of composite products.

Force values of the process are determined in accordance to recommendations cited in Secs. 6.3, 6.4, 7.3, and 7.4 as well as formulas presented below.

8.2 Structure and Properties of the Material Taking Pultrusion Conditions into Account

Physicomechanical characteristics of employed materials along with strength and stiffness of the final article define the pultrusion conditions. The forces arising in the article while clamping in the pulling mechanism may deform and, at certain strength parameters, may break the material.

As an example, let's consider the stress–strain state of a tubular rod fixed in a clamping device, which will help us to determine the corresponding limitations on structural, geometrical, and physicomechanical characteristics of the material as it is pulled in a braiding–pultrusion line.

Because the clamping device is rather long, let us isolate a ring of width b on the loaded part of the rod at a distance from the clamping zone, not considering the edge effect. Thus, following the notions of the construction mechanics of composite structures, the equilibrium equation for a laminated ring under a load spread as shown in Fig. 8.1, is

$$q = q_0 \cos\beta \tag{8.1}$$

wherefrom

$$\begin{cases} N' + \dfrac{Q}{R} + t = 0 \\ M' - Q + m = 0 \\ Q' - \dfrac{N}{R} + n = 0 \end{cases} \tag{8.2}$$

As for the physical relations, they are of the form

$$\begin{cases} N = B\left(V' + \dfrac{W}{R}\right) + C\theta' \\ M = C\left(V' + \dfrac{W}{R}\right) + D\theta' \\ Q = K\left(\theta + W' - \dfrac{V}{R}\right) \end{cases} \tag{8.3}$$

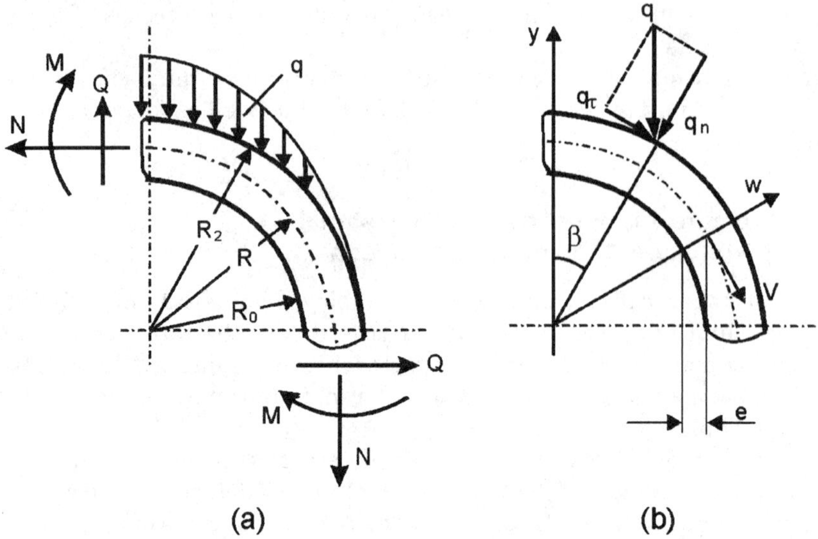

Figure 8.1 Loading scheme.

where N, Q, and M are the longitudinal and transversal forces and the bending moment in the ring section; R is the initial circumference radius; W, V, and θ are the displacements and section turning angle; and B, C, D, and K are the membrane, mixed, flexure, and shear stiffness values of the section, respectively.

The tangential (t), normal (n), and moment (m) loads reduced to a neutral axis while taking expansion of the current load into account are

$$q_{\mathrm{n}} = q_0 \cos^2 \beta$$
$$q_\tau = \frac{q}{2} \sin \beta \tag{8.4}$$

and

$$\mathrm{t} = B\left[1 + \frac{(h-e)}{R}\right] q_\tau$$
$$\mathrm{n} = -B\left[1 + \left(\frac{h-e}{R}\right)\right] q_{\mathrm{n}} \tag{8.5}$$
$$\mathrm{m} = B(h-e)\,[1 + (h-e)R] q_\tau$$

The loading type indicated above is chosen because of the grip design that vertically displaces rubberized lodgments (Fig. 8.2).

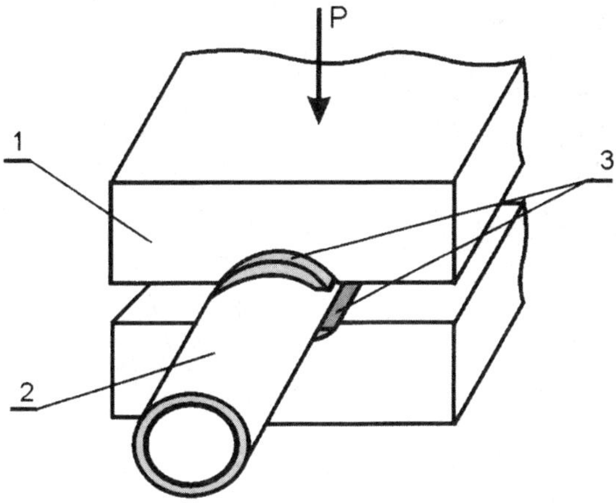

Figure 8.2 Clamping device of the process line: (1) work-holder's body; (2) article being pulled; (3) lodgments.

The total force applied on the pulled rod is

$$P = \int_{-\pi/2}^{\pi/2} g_0 \cos\beta \cdot d\beta \tag{8.6}$$

Under certain e values, the mixed stiffness C turns to zero and

$$\begin{gathered} B = J_0 \qquad D = J_2 - \frac{J_1^2}{J_0} \qquad e = \frac{J_1}{J_0} \\ K = h^2 \left\{ \sum_{i=1}^{4} \frac{2R + r_i + r_{i-1}}{2G\beta_Y^{(i)} R} (r_i - r_{i-1}) \right\}^{-1} \end{gathered} \tag{8.7}$$

where r is the distance to corresponding material layers (Fig. 8.3), and G is the shear modulus of each layer.

Values of J_j are found from

$$J_j = \frac{2R}{j+1} \sum_{i=1}^{4} A_0^{(i)} \frac{r_i^{j+1} - r_{i-1}^{j+1}}{2R + r_i + r_{i-1}} \qquad (j = 0,\ 1,\ 2,\ 3) \tag{8.8}$$

Hence, solution of Eqs. (8.3) and (8.4) for the forces and displacements will be

$$\begin{aligned} N &= N_A \cos\beta - Q_A \sin\beta \\ Q &= Q_A \cos\beta - N_A \sin\beta - Rt \end{aligned} \tag{8.9}$$

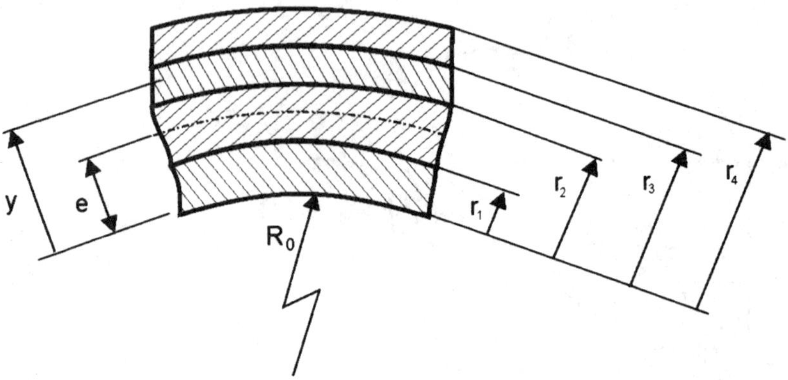

Figure 8.3 Scheme of estimating location of material layers.

$$M = M_0 + RN_0 + RQ_A \sin\beta - RN_A \cos\beta - R\int_0^\beta (m + Rt)d\beta$$

$$\vartheta = V_A \cos\beta - W_A \sin\beta + R\theta_A$$

$$W = RN/B + W_A \cos\beta + V_A \sin\beta$$

$$\theta = \theta_0 + Q(1/B + 1/K) + Rt/B$$

where N_0, Q_0, and M_0 are determined under $\beta = 0$

$$N_0 = 0; \qquad Q_0 = -RPb; \qquad M_0 = -0.116PR$$

$$N_A = \frac{P}{2}\left[1 + \frac{(h-e)}{R}\right]\left(\frac{\cos^3\beta}{3} - 2\cos\beta + \frac{1}{3}\right)$$

$$Q_A = -P\left\{\frac{1}{2}\left[1 + \frac{(h-e)}{R}\right]\left(\frac{\cos^3\beta}{3} - 2\cos\beta + \frac{1}{3}\right) + 1\right\}$$

$$\theta_A = \theta_0 + Q\left(\frac{1}{B} + \frac{1}{K}\right) - \frac{Rt_0}{B} + \int_0^\beta \left[\frac{RM}{D} - (N - nR)\frac{tR^2}{B}\right] d\beta \quad (8.10)$$

$$W_A = W_0 + \frac{R^2}{D}\int_0^\beta M \sin\beta\, d\beta$$

$$V_A = \vartheta_0 - R\theta_0 - \frac{R^2}{D}\int_0^\beta M\cos\beta\, d\beta$$

Values θ_0, W_0, and ϑ_0 are found from boundary conditions

$$\beta = 0 \rightarrow \theta_0 = 0 \qquad \vartheta_0 = 0$$
$$\theta = \frac{\pi}{2} \rightarrow \theta = 0 \qquad \vartheta = 0 \quad (8.11)$$

Because the tubular rod can be broken in points $\beta = \pm\pi/2$ due to certain loading conditions, let us calculate the inner force factors

$$N_{\mathrm{P}} = P\left[\frac{1+\frac{(h-e)}{R}}{6}+1\right]$$
$$Q_{\mathrm{P}} = P\frac{1+(h-e)R}{6} \qquad (8.12)$$
$$M_{\mathrm{P}} = -0.116PR - RN_{\mathrm{P}}$$

Proceeding from Eq. (8.11), the vertical shift of the grips can be found from the flexure at point $\beta = 0$.

$$W_0 = \frac{\pi MR^2}{2D} \qquad (8.13)$$

Stresses that operate in the grip plane can be calculated by the following relations:

$$\sigma_\beta = \frac{E_\theta}{1+\frac{y}{R}}\left(\frac{N}{B}+y\frac{M}{D}\right)$$
$$\tau = \frac{QbE_\theta}{b\left[1+\frac{y}{R}\right]^2}\left[\frac{y+e}{BR}-\frac{y^2-e^2}{D}\right] \qquad (8.14)$$

$$\sigma_{\mathrm{r}} = \frac{E_\theta}{b(1+\frac{y}{R})^2}\left\{\frac{Nb(y+e)}{BR}-\frac{Nb}{R\left(1+\frac{y}{R}\right)}\left[\frac{(y+e)^2}{2BR}+\frac{e^2(2e+3y)-y^3}{3D}\right]\right.$$
$$\left.+\frac{\left[1+\frac{(h-e)}{R}\right]Pb\,(y+e)^2}{16BR\left(1+\frac{y}{R}\right)}-\frac{(h-e)\left[\frac{1+(h-e)}{R}\right]Pb}{24RD\left(1+\frac{y}{R}\right)}[y^3-e^2(3y+2e)]\right\}$$

where E_θ is the elasticity modulus and y is a coordinate at a random point of the section (Fig. 8.3).

The distribution of the normal and tangential stresses with section height is shown in Figs. 8.4, 8.5, and 8.6 as a function of the reinforcement angle of the braided layer.

Analysis of maximum stresses in a section (Fig. 8.7) shows that the most efficient reinforcement of the braided layers is at $\alpha = 45°$. This helps to significantly lower the clamping force of the rod at pulling. Calculation results of deformations in point $\beta = 0$ versus the reinforcement

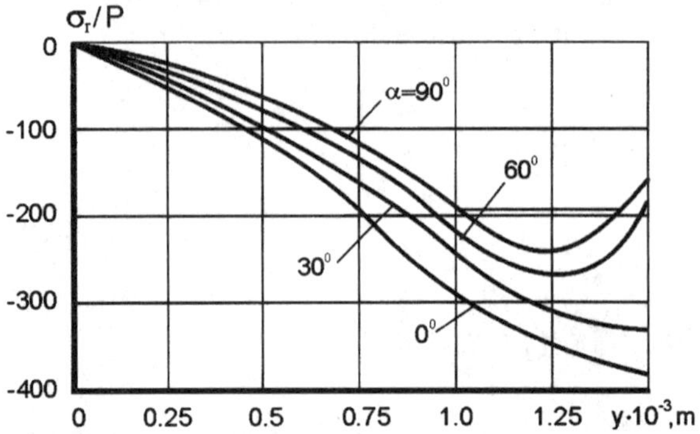

Figure 8.4 Variations in radial stresses with section height as a function of the reinforcement angle of the braided layer.

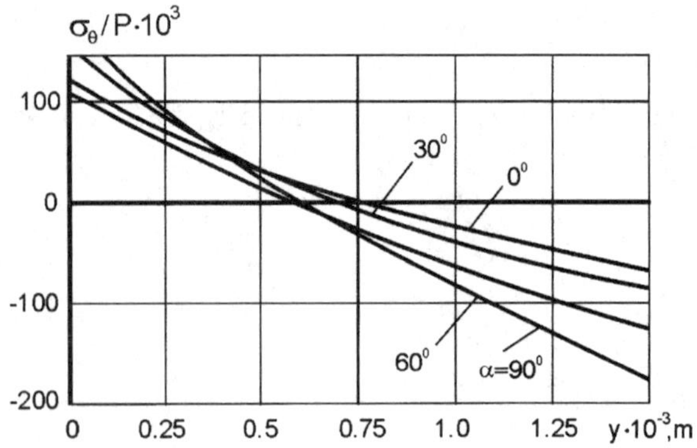

Figure 8.5 Effect of reinforcement angles of the braided layer on tangential stresses with section height.

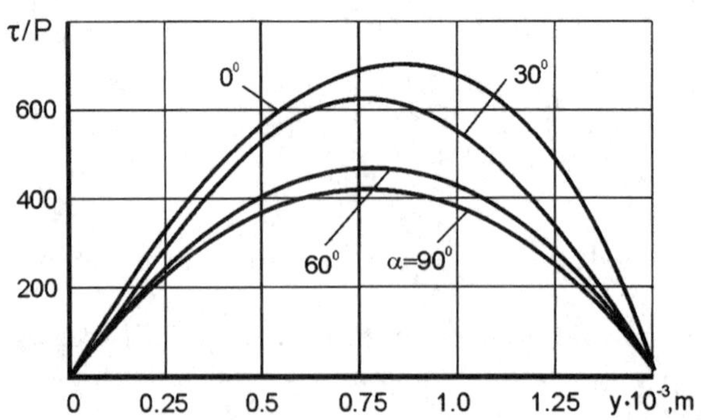

Figure 8.6 Distribution of tangential stresses over the section height for different reinforcement angles of the braided layer.

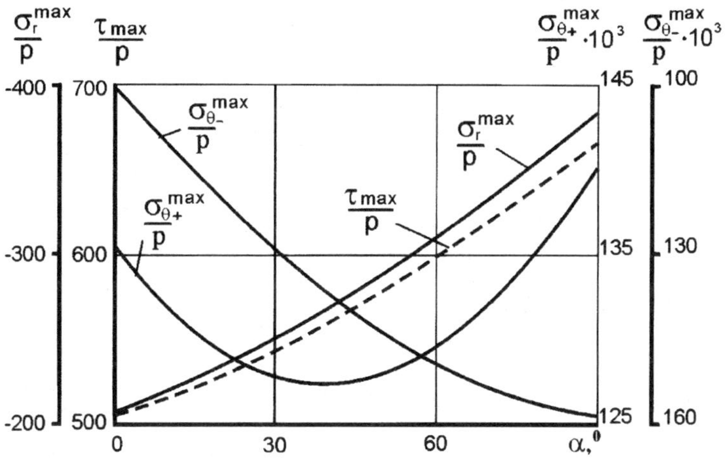

Figure 8.7 Effect of reinforcement angles of the outer layers, where $h_3/h_4 = 1$, on the maximum stresses in the material structure.

angle of the braided layers and different transversal-to-longitudinal reinforcement ratios are presented in Fig. 8.8. On the basis of the data obtained, we can calculate the minimal gaps in the clamping mechanism so as to make the clamping conditions most efficient.

The limiting clamping force has been defined, taking into account the structural and physicomechanical characteristics as well as the strength criterion, for which purpose stresses in each layer were found. Because the limiting clamping force should not cause the material to crack, the strength condition has been taken for each layer in the

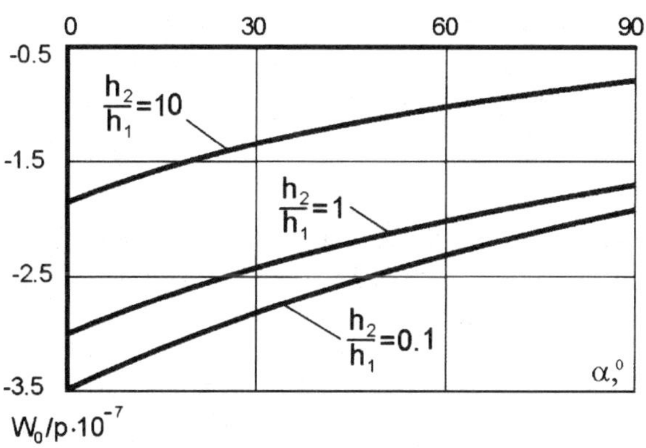

Figure 8.8 Effect of reinforcement angle of the braided layer and transversal-to-longitudinal layer ratios on vertical shift at point $\beta = 0$.

form [325]

$$\sigma_1^2 d_1 + 2\sigma_1\sigma_2 d_2 + \sigma_2^2 d_3 = \frac{\Delta^2}{h^2} \tag{8.15}$$

where

$$d_1 = \left(1 - \nu_C^2\right)\bar{\sigma}_r^2 \frac{a_2^2}{\left(R_C^+\right)^2} + \frac{a_3^2}{T_C^2}$$

$$d_2 = \left(1 - \nu_C^2\right)\bar{\sigma}_r^2 \frac{a_2 b_2}{\left(R_C^+\right)^2} + \frac{a_3 b_3}{T_C^2}$$

$$d_3 = \left(1 - \nu_C^2\right)\bar{\sigma}_r^2 \frac{b_2^2}{\left(R_C^+\right)^2} + \frac{b_3^2}{T_C^2}$$

$$\Delta = Q_{11}Q_{22}Q_{66} - Q_{12}^2 Q_{66}$$

The coefficients a_2, b_2, a_3, and b_3 and stress concentration factors $\bar{\sigma}_r$ have been calculated elsewhere[325] R_C^+ and T_C are the tensile and shear strength values of the binder, respectively. The least of the values have been accepted.

It is evident that, in order to attain a needed pulling force (see Chap. 5), the corresponding clamping force must to be created for the article being pulled in the grips of the clamping device:

$$P_Z = \frac{b}{L_f} \times \frac{P}{f} \tag{8.16}$$

where L_f is the grip length.

According to the theoretical analysis, only the materials with certain structural and physicomechanical characteristics can be pultruded. To verify the dependencies obtained, the effect of the longitudinal-to-transversal layer ratio and the reinforcement angle on the bearing capacity of tubular rods at cross-reduction has been experimentally studied.

The lodgments dismantled from the process line have been used in tooling during the tests. The force corresponding to initial crack generation, displacing the active grip of the testing machine and occurring together with deformation of the tubular rod in the joint plane has been determined. Tubular samples were tested after thermal treatment and prior to clamping. It is apparent that, along with the operating load, force P_F of crack initiation depends on the grip length as well. Grips with removable lodgments were used to clarify the character of P_F dependence on the length of applied load. Note that tubular rods had

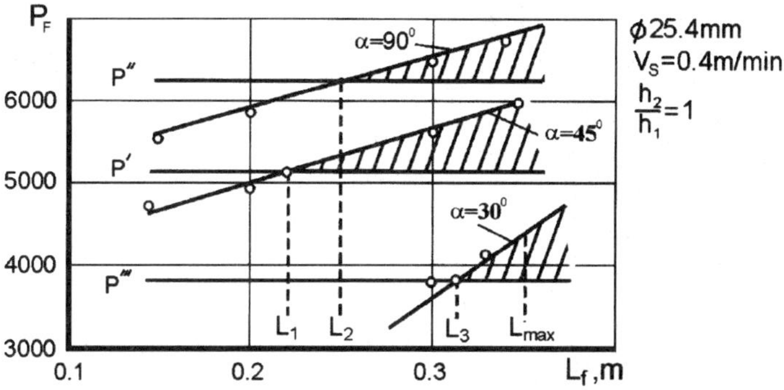

Figure 8.9 Effect of lodgment length and reinforcement angle on the load at which continuity is lost at cross-reduction; L_{max} is the largest efficient length of the lodgment; L_i is the critical length of the grips; and P', P'', and P''' are the required pulling forces.

different structural parameters. Experimental data illustrating variations of tubular, 25.4-mm diameter rods are presented in Figs. 8.9 and 8.10. The values indicated therein by dotted lines represent the pulling force with account of the friction coefficient of the glass plastic–rubber pair (f_{fr} = 0.58 according to Ref. 201). As seen in the figures, pulling without disturbing the continuity of the material is possible only if the grip length exceeds some limiting value L_{lim}. In a number of cases, the design features of the pulling device cannot ensure the required length of the grips, and such profiles cannot be drawn.

With increasing size of the pultruded article, the force corresponding to crack onset also grows (Fig. 8.11).

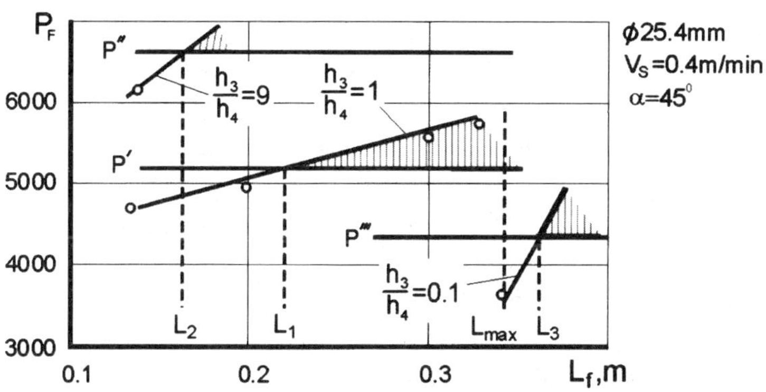

Figure 8.10 Effect of grip length and ratio of longitudinal to transversal layers on the force leading to rupture at lateral reduction.

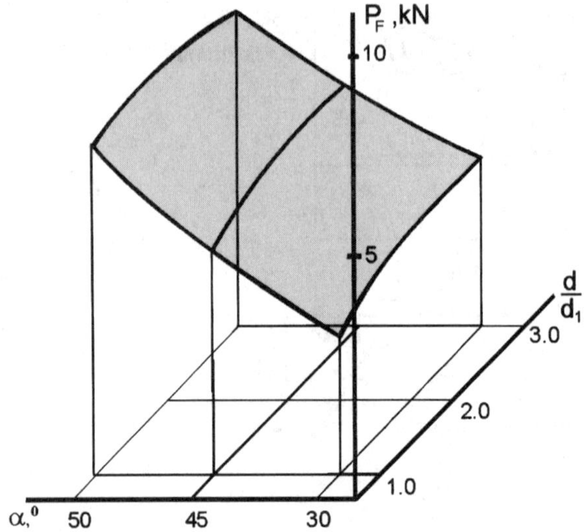

Figure 8.11 Effect of reinforcement angle of the braided layer and article dimensions on the force corresponding to rupture onset.

When estimating deformation of the tubular rod, one should consider the presence of rubber liners on the lodgment inner surface. Therefore, shift of the grip, except for W_0, includes deformation of the rubber, which is

$$\Delta = \frac{2P}{3\pi L_f G} \times \frac{L_f^2 + 3(R_O + R_I)^2}{L_f^2 + 6(R_I - R_O)^2} \times \frac{(R_I - R_O)^3}{(R_I + R_O)^3} \tag{8.17}$$

where R_O is the outside diameter of the article, R_I is the inner grip length, and G is the shear modulus of rubber ($G = 0.5$ MPa).

Test results of the rods with different structural characteristics of their material are presented in Table 8.1. The rods chosen were 25.4 mm

TABLE 8.1 Test Results of Tubular Rods under Cross-Reduction (Ø 25.4 mm, $L_f = 350$ mm)

			P_F^E, N		Grip shift, mm		
h_3/h^4	α°, deg	P_F^T, N	Value	ν,%	$(W_0+\Delta)_T$	$(W_0+\Delta)_E$	ν, %
0.1	30	3320	3300	11.6	0.447	0.71	7.2
	45	3430	3650	8.4	0.301	0.55	4.3
	60	3810	3960	10.3	0.292	0.25	5.7
1.0	30	4270	4050	12.1	0.352	0.50	4.2
	45	4565	5400	10.7	0.339	0.41	6.0
	60	5300	6450	11.1	0.283	0.32	5.3

in diameter and with wall thickness of 1.5 mm. Deviation of theoretical results P_F^T from experimental values P_F^E is less than 20.7 percent. Shifting of the active grip measured during testing corroborated with theoretical data, although their deviation values are by far higher. This is attributed to deformations of the lodgment itself not considered in the experiment.

The derived theoretical and experimental results form the basis for recommendations on the choice of the material and design of the clamping mechanism promoting high-quality pultrusion.

8.3 Tooling and Equipment Chosen for the Process

The functional diagram equipment for continuous production of complex-profile wares can be based on a typical scheme of pultrusion of named products from unidirectional composites. It should also include some special-purpose equipment described earlier. The scheme illustrated in Fig. 8.12 is recommended for application in production of complex-profile composite articles.

8.3.1 Strainers

The reinforcement is reeled off the bobbins or a reel placed on carriers at slight braking and is collected into strands, after which it is then directed into an impregnator. A bobbin carrier consists of a metal frame with horizontal axes. For installation of a large-size bobbin,

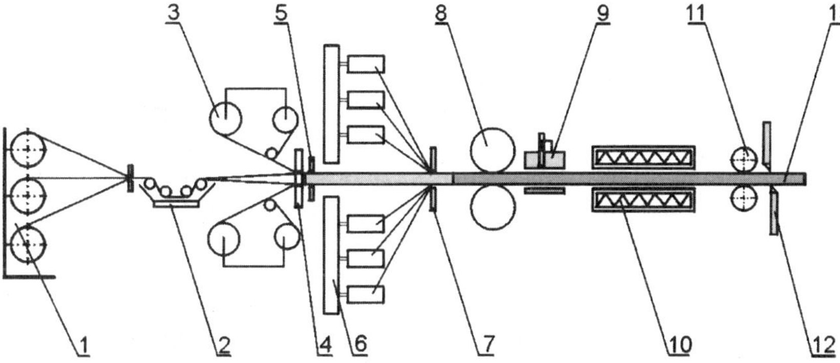

Figure 8.12 Functional diagram for the continuous production process of complex-profile wares of force purposes: (1) bobbin carrier with reinforcement for profile flanges; (2) impregnator; (3) bobbins with prepreg for post molding; (4) distributor; (5) preforming die for the semi; (6) braider; (7) calibrating die; (8) preforming die for profile post; (9) mandrel unit; (10) heated die; (11) draw-out device; (12) cutter; (13) final article.

Figure 8.13 Rack-type bobbin carrier showing pretension system with longitudinal reinforcement.

e.g. with glass fibers, rack-type carriers are used, where the reinforcement is reeled off inside the carriers (Fig. 8.13). Their size depends on the amount of braids necessary for certain dimensions of articles.

The reinforcing material is tightened either by braking of the bobbins or by the reinforcement itself. Methods of braking the fibers or braids applied in low-inertia and inertia-free mechanisms of tension regulation have been considered elsewhere.[54,105] Among them is a frictional method that creates tension force via friction of a braid as it passes between felt plates that are pressed against each other. This method is, however, inapplicable for fine-fibrous, easily damaged fillers.

In a number of strainers, the reinforcement is tightened by slightly braking bobbins fit with either disc or band, electromagnetic or pneumatic brakes. Pneumatic and hydrodynamic methods of tightening that combine processes of tension with impregnation have recently become more popular.[120]

Still, the most simple and efficient method seems to be the one based on passing reinforcing braids through a system of braking, rotating rollers. The braking force of each roller should be less than the friction force that ensures motion of the braids together with the roller without sliding. The friction force value depends on the wrap angle of the roller by a fibrous material and on the pretension force arising from braking of the bobbins (see Fig. 8.14).

Figure 8.14 Pretension reinforcement device.

8.3.2 Impregnation–squeezing devices

The fibrous filler is impregnated with a binder as the strands of fibers are drawn through special impregnation baths where the binder is heated to 60°C to lower its viscosity and improve wetting conditions.

Sometimes, easily damaged materials (some types of carbon braids) for which strong bending is unacceptable (boron or glass fibers) are impregnated using the method of immersed rollers, in which the braids are placed on the upper part of the rotating roller while the lower part of the rotating roller is immersed in the bath with the binder. When rotating, the roller captures the binder, which is brought into contact with the fibrous material. Such fillers can be also impregnated by methods that combine hydrodynamic impregnation with tension.

In other methods, the braids are directed immediately into the impregnation bath through a system of free rollers or a guiding comb. To prolong the time of contact and take full advantage of hydrodynamic effects, the method of multiple bending of the fibrous reinforcement around a system of revolving or immovable rollers in the impregnation bath is used (Fig. 8.15).

Upon impregnation, excess of the binder is squeezed out when the braids pass through a system of wringer rolls or filters. The braids are wetted first to prevent soiling of the equipment during further processing.

8.3.3 Distribution devices

After impregnation and squeezing, the reinforcing material is fed into a distributor made as a ring with a system of orifices or an annular comb

Figure 8.15 Impregnation–squeezing device.

fit with a system of guiding bars. The required quantity of reinforcement braids is spread over the mandrel surface in each zone by a distributor.

In case of complex-profile articles, preimpregnated fabric, or "prepreg," is fed together with the flange material into the distributor for molding of the profile post because polyethylene films that separate the prepreg layers are peeled off prior to entering the distributor and reeled onto a special bobbin.

After the distributor, the impregnated reinforcement laid on the mandrel is then fed into the preforming die. The configuration and dimensions of the mandrel and die should comply well to allow for a given correlation of the reinforcement and binder in each cross-section area of the semifinished product.

8.3.4 Calculation of preforming die dimensions

For the calculations, it is necessary to define cross-sectional areas of each structural element of the complex-profile article. Their values can be found from the following relations.

Area of the main material of the flange

$$F_{\mathrm{s}} = b_{\mathrm{s}}\delta_{\mathrm{s}} \tag{8.18}$$

Areas of the post and shoulders occupied by embedded elements:

$$F^{*}_{\mathrm{me}} = (b_{\mathrm{r}} - 2\delta_{\mathrm{b}})\left[H + \frac{B - b_{\mathrm{r}}}{2} - 2(\delta_{\mathrm{s}} + \delta_{\mathrm{b}})\right] \tag{8.19}$$

Area of the braiding

$$F_b = (\Pi - 4\delta_b)\delta_b \quad (8.20)$$

Perimeter of the profile article

$$\Pi = 2(H + 2B - b_r) \quad (8.21)$$

As indicated in Sec. 6.2, the filling degree of each material with fibrous reinforcement is conditioned by the type of the article and process regimes.

To attain the needed filling degree of each profile element, one should calculate its perimeter without the braiding layer by the following formula:

$$\Pi^* = 2(H + 2B - b_r - 4\delta_b) = \Pi - 8\delta_b \quad (8.22)$$

As for the preforming die, its rated diameter should be equal to

$$D_C = \frac{\Pi^*}{\pi} \quad (8.23)$$

And yet, the saturated reinforcement should contain some excess of the binder to further impregnate the braiding layer.

The required additional area of the binder is determined by Eq. (6.2):

$$F_{b1}^* = F_{b1}(1 - \varphi_{b1}) \quad (8.24)$$

where φ_{b1} is found from Eq. (6.1).

The actual diameter of the preforming die can then be calculated as follows:

$$D_{pm} = \sqrt{\frac{4F_{b1}^*}{\pi} + D_C^2} = \sqrt{\frac{4\delta_b}{\pi}(\Pi - 4\delta_b)(1 - \varphi_{b1}) + D_C^2} \quad (8.25)$$

Die is shown in Fig. 8.16.

8.3.5 Estimation of geometrical parameters of the mandrel

The mandrel covered by the flanges and post is calculated so that its cross-sectional area together with that of the flanges, post, and shoulders fills completely that of the preforming die:

$$F_{pm}^C = F_m + 2F_s + F_{me}^* \quad (8.26)$$

where $F_{pm}^C = \pi D_C^2/4$.

Figure 8.16 Calibrating die.

To meet this condition, the mandrel should have a peculiar cross-sectional shape (Fig. 8.17).

Let us presume that the wetted section of the parts filled with the impregnated reinforcement presents an intersection of the shaping circle of the die of rated radius $r_C = D_C/2$ and corresponding supporting circle of the mandrel R_{br}^{m} or R_{br}^{s} (Fig. 8.18).

The overall length of the post together with the shoulders will be

$$L_{max}^{r} = H + B - 2\delta_s - 4\delta_b \tag{8.27}$$

If this layer is laid on the outside surface of a semi, its length will define the location of intersection points on circles A and B and their coordinates z_A and y_A.

$$L_{max}^{r} = r_C\alpha \tag{8.28}$$

$\alpha = L_{max}^{r}/r_C$ is the angular size of the arc AB on the circle of radius r_C.

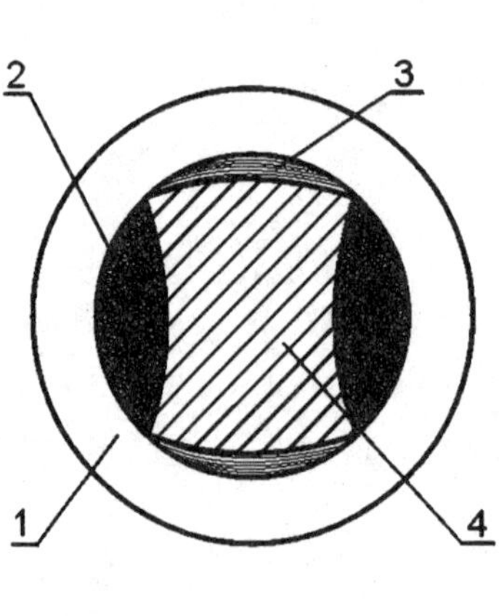

Figure 8.17 Distribution of reinforcement over the mandrel surface at preforming stage: (1) preforming die; (2) flange material; (3) post material; (4) mandrel.

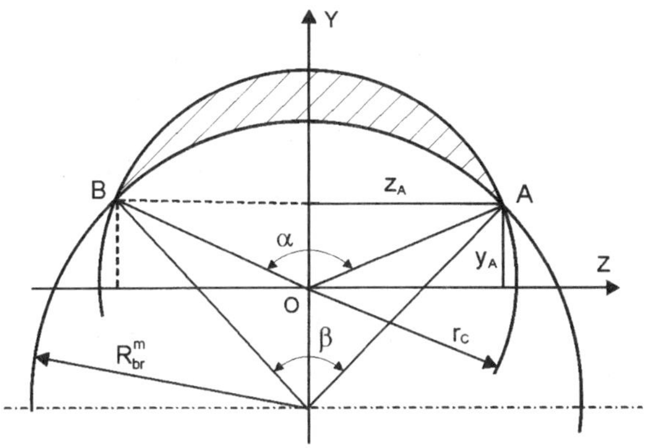

Figure 8.18 Calculation scheme of cross-sectional shape of the mandrel.

The coordinates of points A and B can be expressed simply as parameters of the forming die or of the supporting surface of the mandrel.

In the former case, the system of equations defining these coordinates will be

$$\begin{aligned} z_A &= r_C \sin(\alpha/2) \\ y_A &= r_C \cos(\alpha/2) \end{aligned} \tag{8.29}$$

These coordinates for the supporting circle of the mandrel will be written in terms of the mandrel parameters to be determined.

$$\begin{aligned} x_A &= R_{br}{}^{m} \sin(\beta/2) \\ y_A &= R_{br}{}^{m} \cos(\beta/2) - y_0 \end{aligned} \tag{8.30}$$

where $R_{br}{}^{m}$ is the radius of the mandrel in the zone where the post material is laid; β is the angular size of arc AB on the circle of radius $R_{br}{}^{m}$; and y_0 is the center-to-center distance of circles r_C and $R_{br}{}^{m}$.

An unambiguous solution can be based on the condition of the equality of the material and gap areas between the mandrel and the die.

$$F^*_{me} = r_C^2(\alpha - \sin\alpha) \pm \left(R_{br}{}^{m}\right)^2(\beta - \sin\beta) \tag{8.31}$$

Similarly, mandrel parameters can be calculated in the zone where the main material of the flange is laid.

Prior to the step above, the following condition is checked:

$$F^*_{me} \geq \frac{r^2}{2}(\alpha - \sin\alpha) \tag{8.32}$$

If this condition is obeyed, a minus sign is used in Eq. (8.31); if not obeyed, a plus sign is used. The system of equations (8.28) through (8.31) has no analytical solution but can be easily solved by numerical methods.

If a layer of overall length is laid directly on the mandrel surface, Eq. (8.28) should be substituted in the system by the following equation:

$$L^{\mathrm{r}}_{\mathrm{max}} = R_{\mathrm{br}}{}^{\mathrm{m}}\,\beta \tag{8.33}$$

For the value of $L^{\mathrm{s}}_{\mathrm{max}}$ in calculations of the mandrel section shape, we should take the value from the following formula:

$$L^{\mathrm{s}}_{\mathrm{max}} = \pi r_{\mathrm{C}} - L^{\mathrm{r}}_{\mathrm{max}} \tag{8.34}$$

When a fabric prepreg is used for the post and shoulders, it should incorporate an amount of the binder that can ensure a needed filling degree. So the entire binder volume for impregnation of the braided layer is initially included into the volume of the flanges and is then redistributed over the perimeter of the semi during further molding stages.

Note that a given filling degree is commonly realized at the final stages. At intermediate stages, the semi usually includes air and gas inclusions, solvents and diluters, initiating agents, and other ingredients. Thus it is recommended that the rated volumes of materials be increased by roughly 5 percent at the initial and intermediate stages of force treatment.

8.3.6 Design solutions of braider elements

The primary requirements imposed on the braider elements are (i) undamaged passage of the reinforcement through the strainers and (ii) stable tension. Currently used braiding equipment used in production of cables and wires for screens and protective covers can be subdivided by their kinematic properties into shuttle and rotary types. Both are intended for 2D networks.[153,428]

Fiber carriers in shuttle machines move over circular undulation trajectories around the braiding center. There are two types of fiber carriers: some move clockwise, others move counter-clockwise. In rotary machines, two groups of reels with fibrous materials rotate toward each other over two concentric circles about the article axis. Different mechanisms exist for obtaining interweaving (driving and planetary) that shift strands of the outer row under and over those of the inner row.

Comparison of shuttle and rotary machines has shown that the latter are faster while the former are more compact and simpler in structure. In addition, it is possible to construct a device that is based on the design of the shuttle machine, exercising 3D braiding, by involving fiber carriers from the third direction. To do this, the shafts of the driving gears are made hollow and are used as loopers for the fibers of the third direction. It is much more difficult to make these modifications in a rotary machine.

Analysis of existing braiding devices has shown that the device most acceptable for the basic variant is the shuttle machine, whose fiber carriers ensure qualitative processing of reinforcement materials.

The chief design peculiarity of the braider is likely to be the type of the joint separating the spindles of the left and right directions. Most varied are the schemes that separate by a system of milled grooves in the lead plate of machine over which the spindle shuttle moves. This, however, necessitates a complex and costly base part.

An experimental braider design intended for carbon fibers without a lead plate with milled guides has been proposed. For separating the spindles, a pointer mechanism has been used. According to the scheme shown in Fig. 8.19, the braider contains a base plate (1) parallel to which two-level guides (2, 3) are installed. In certain places of the plate, hollow shaft-loopers (4) are mounted driven by pinions (5, 6) that close a kinematic circuit. Plates (7) and supporting grips (8) are connected with shafts (4). The lower part of spindles (9) is a shuttle made as a pair of rollers (10) seated in an axis (11). Pointers (12) are installed in pairs in holes of guides (2, 3) and are connected to a rod (15). Angular displacement of the pointers is restricted by rests (14).

Spindles (9) are actuated during operation by grips (8). As the spindle approaches the separation zone, rollers (10) run over point (12) and

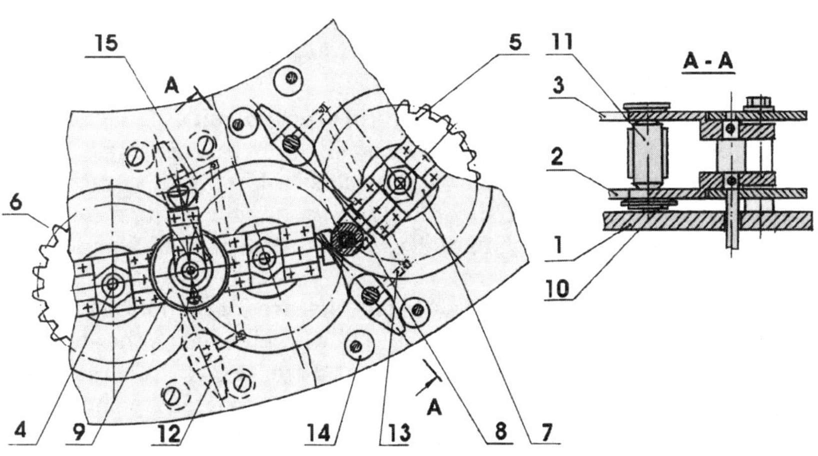

Figure 8.19 Design of a braider with pointer separation mechanism of spindles (described in detail in the text).

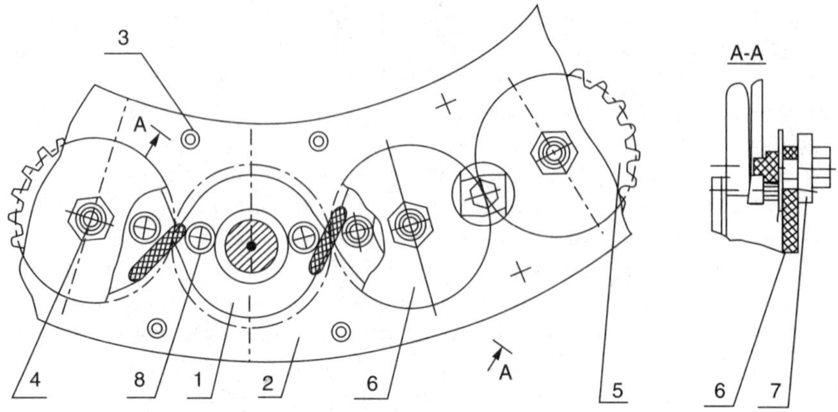

Figure 8.20 Driving gear design (described in detail in the text).

move it to the initial position of the transfer trajectory. The moved point connected with rod (13) transfers it in the position of continuation of the transfer trajectory. The spindles then move the points along until they are captured by the next grip and transferred to the opposite guide.

In this design, it is important that the spindles are kept in a stable position and that the braids are tightened to a required value during braiding.

Drawbacks of this design include its intricacy and the necessity of regulating the switching mechanism as well as the high velocity of the spindles.

A simplified modification of this structure with profiled guiding slots in the base plate is illustrated in Fig. 8.20. It is fit with shut-off disc mechanisms and simplified, overlaid guides. This design is used in textile and cable industries and can be employed for braiding composite materials as well.

The braider contains a base plate (1) and overlaid guides (2, 3). Hollow shafts operating like loopers (4) are installed on the bearings on the plate and are actuated by gears (5) that close a kinematic circuit. The shafts are rigidly linked with discs (6) having slots for conjugation with spindle shuttles (7). Shut-off discs (8) are mounted in transfer points of the spindles.

During operation, shuttles (7) with spindles are set in motion by discs (6). The undulation trajectory of the spindles is achieved as the shuttles slide over grooves in guides (2, 3). Radial displacement of shuttle (7) relative to the center of rotation of disc (6) is restricted by the depth of slots in disc (6). As shuttles (7) approach the transfer point, they come into contact with shut-off disc (8) and pass over into the next disc slot.

Figure 8.21 General view of distributor drive.

The design described is simple in production and develops higher velocities compared to the previous one, although its spindle stability is somewhat worse. A general view of a distributor drive is shown in Fig. 8.21.

In conditions of experimental production, it is possible to use a universal lathe as a braider drive and to exercise driving from a spindle as the article is displaced in the braiding center by the lathe carriage.

The main unit of a braider upon which the fiber quality depends is a spindle. Spindles of commercially produced braiders represent movable units carrying changeover fiber carriers (reels or bobbins) and are intended for processing glass and organic fibers as well as low-modular carbon fibers. These designs are, in fact, incapable of processing high-modular carbon fibers. The threading track of a typical spindle presents a line with great many small-radius bends (5–10 mm). Passage of high-modular fibers through this line results inevitably in fluffing and rupture. As investigations have shown, for qualitative processing of high-modular carbon fibers, it is necessary to strictly observe the limiting bend radius of the fibers (above 30 mm) and minimize the contact length with the movable units.

Special spindle structures for braiders have been developed with axial flow of fibers and pointer scheme of spindle separation (Fig. 8.22).

The spindle contains a base stem (1) on which a shuttle is mounted made up of a pair of rollers (2), distance sleeve (3), and pin (4). Stem (1) is rigidly fixed by a platter (5) with a friction lining (6). Bushing (8) is pressed against the platter by a nut (7), on which the bushing

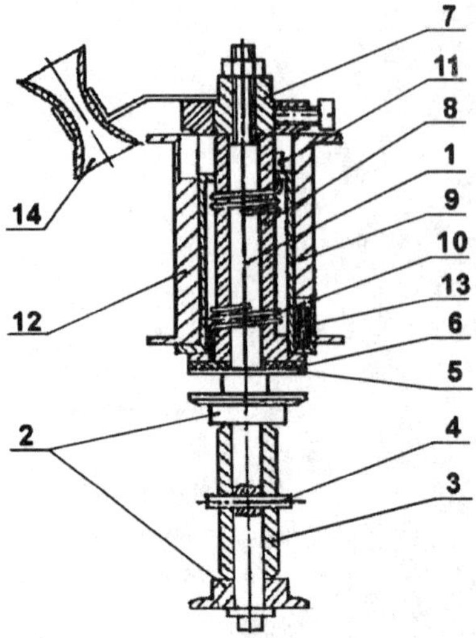

Figure 8.22 Spindle design for a braider with pointer separation (described in detail in the text).

cylinder (9) is installed and connected with it through a torsion spring (10). Rotation of the cylinder around the bushing is restricted by a stop (11). A bobbin with a fibrous material (12) is mounted on cylinder (9) and is fixed by dowels (13). Fibers flow from the bobbin through the die (14).

During operation of the braider, the spindle moves about the braiding center and the fibers are reeled off the bobbin (12). The bobbin rotates together with cylinder (9), which is accompanied by winding of a spring (10). Upon making a turn, the cylinder contacts the stop (4) and then starts to rotate jointly with bushing (8). Rotation of bushing (8) is hampered by a braking torque appearing in the friction lining (6). Tightening force of the spring generates the braking torque and induces tension force on the fibers. By changing the distance between the spindle and braiding center (owing to the undulating trajectory of spindle motion), the fibers are pulled tight by spring (10). Platter (5) perceives the braking torque through the pin that is regulated by the nut (7).

A shortcoming of this design is attrition of fibers as a result of axial flow of threads during reeling, which can be related also to its advantage because the dimensions of the bobbins and, consequently, their capacity become unlimited.

The described spindle with a radial fiber flow is intended for braiders with shut-off discs for spindle separation (Fig. 8.23).

Such a spindle contains a base bracket (1) with a shuttle (2) fixed on its lower part and a guiding die (4) on its upper plank (3). Roller (5) is fixed by a pin (6) in a hole of the bracket edge. Roller (5) is connected

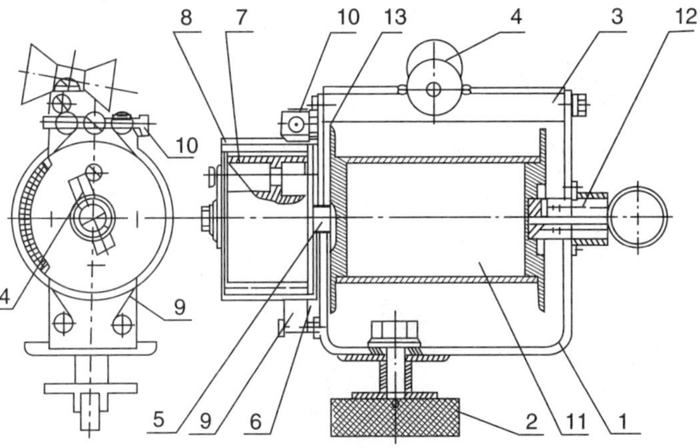

Figure 8.23 Spindle of a braider with shut-off discs (described in detail in the text).

through torsion spring (7) with cylinder (8) interacting with a brake (9) that is regulated by a screw (10). Reel (11) is fixed by a spring rest (12) and a pin (6). Plank (13) is rigidly connected with the roller (5), whose rotation about cylinder (8) is restricted by stop (14).

During operation of the braider, reel (11) revolves under the action of the reeled fibers. The reel rotates together with roller (5), which goes hand-in-hand with winding of the spring (7). Upon making a turn, the roller (5) effects the stop (14) of cylinder (8) through plank (13). Further on, roller (5) rotates together with cylinder (8). The tension force of the fibers is dependent on tightening of brake (9) by the screw (10). The forming slack of the thread is removed by spring (7).

The spindle described is intended for braiding carbon fibers, during which their damageability is minimized and velocities are rather high. A general view of the spindle is shown in Fig. 8.24.

The spindle design described above has been tested in field conditions, and it showed high tension force stability of the reinforcement and failure-free operation. A general view of a braider with described spindles is presented in Fig. 8.25.

A braider with cross-winding of reinforcement materials is shown in Fig. 8.26.

8.3.7 Design of the calibrating die location

The semifinished product is calibrated directly after the braiding layer packing. For this purpose, the calibrating die is installed in the section where the one-sheet hyperboloid of the free braiding surface intersects with the cylindrical surface of the semi.

Figure 8.24 General view of a spindle.

Because the reinforcing braids are arranged in a line that is tangential to the mandrel surface and at angle α to its long axis, the following relation is to be obeyed (Fig. 8.27):

$$\tan\alpha = \frac{y}{L_C} \tag{8.35}$$

where y is the projection of the free length of the winding braid to the bobbin plane and L_C is the distance from the bobbin plane to calibrating die place.

Figure 8.25 Braiding device.

Figure 8.26 Cross-winding braider.

If a braid is always in a plane perpendicular to the semi radius taken at the point of its contact with the winding braid, then value y can be calculated as follows:

$$y = \sqrt{R_{\mathrm{m}}^2 - \frac{D_{\mathrm{C}}^2}{4}} = \frac{1}{2}\sqrt{4R_{\mathrm{m}}^2 - D_{\mathrm{C}}^2} \tag{8.36}$$

where R_{m} is the design value of the braider defining rotation centers of transmitting mechanisms or a mean radius of the motion trajectory of the bobbin with reinforcement braids.

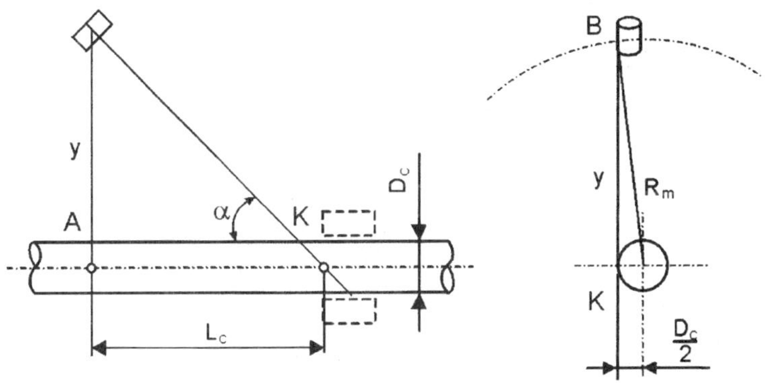

Figure 8.27 Calculation scheme of the calibrating die position.

Figure 8.28 Calibrating die position.

From Eq. (8.35), the calibrating die position will then be defined as follows:

$$L_C = \frac{\sqrt{4R_m^2 - D_C^2}}{2\tan\alpha} \tag{8.37}$$

The position of the calibrating die in a process line is shown in Fig. 8.28. In Fig. 8.29 is shown the position of a preheating device of the semi before it enters the die.

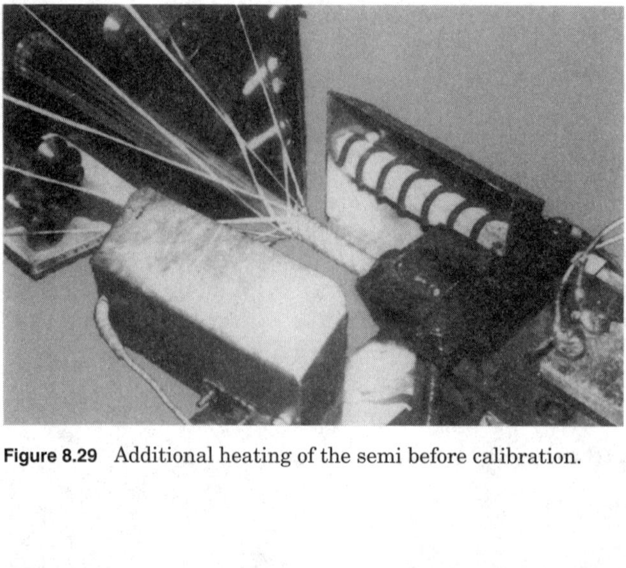

Figure 8.29 Additional heating of the semi before calibration.

8.3.8 Preforming mechanism of the profile post

Analysis of the preforming process of the profile post proves that the requirements of centering the semi, namely, separation of flange materials and the profile post should be fulfilled during forming and joining of parts of the post at different portions of the mandrel surface. For this purpose, two elastic plates are installed at the mandrel outlet, whose width is calculated so that the area occupied by the flange material remains constant at each molding moment (Fig. 8.30). The plates separate the flange material from the post molding zone at all stages of its shaping.

Molding is realized either by the forming plates or by the molding rollers. The curvature radius of the forming elements and their location coordinates are related through Eq. (6.24) and can be defined under specified conditions by a complex process and structural parameters of the complex-profile article.

8.3.9 Stitching process

Both standard and specifically designed equipment can be used for stitching. Sewing machines are preferable from the viewpoint of their universal availability and high-quality factory assembly. Their negative feature is the necessity of adjusting the time when the needle is in the semi with that when the whole complex or its part is moving.

The velocity of the forming profile, in practice, is within limits of 0.2–0.6 m/min, while at standard stitching, the time the needle is in

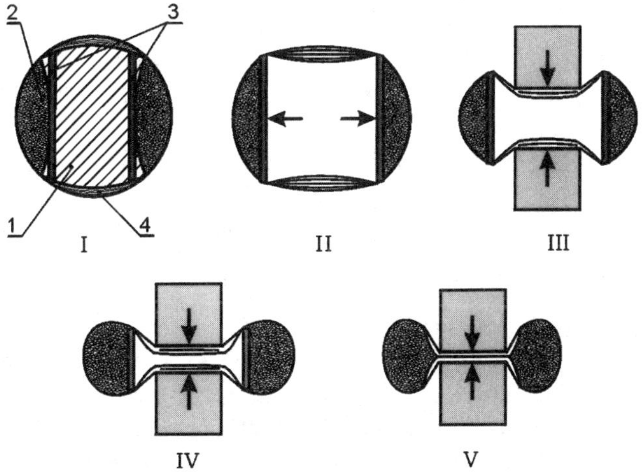

Figure 8.30 Preforming stages of profile post: (1) mandrel; (2) flange material; (3) separating plates; (4) post material.

Figure 8.31 Pneumatic stitching machine.

the material relates to the time it is out of the material at a 1:1 ratio. Therefore, if profile articles are of small size and the stitch length is 2–4 mm, it is expedient to use standard or modified sewing equipment.

At the same time, in case of large-sized profile articles and long stitches, the low velocity of the formed material supports using pulse methods with low damageability of the semi. Pulse regimes for inserting the needle are realized by electromagnetic, hydraulic, or pneumatic systems. The time the needle stays in the material can be reduced and be limited by only strength parameters of the sewing material as it follows from Eqs. (7.47), (7.49), and (7.50), due to strong effects of dynamic factors. One type of pneumatic machine used for stitching reinforced polymer composites is shown in Fig. 8.31.

8.3.10 Devices for complete molding of profiles

The forming die presents a collapsible metal mold whose working surfaces are polished and chrome-plated. The die length depends on the pulled profile dimensions, binder type, and process rate and is usually within the 300–1500 mm range. The working channel acquires a geometrical shape of the profile formed and has parallel walls with rounded edges to avert fiber fracture.[114,164,317]

The die is commonly heated by flat electric heaters mounted inside the die. To increase its efficiency, high-frequency induction heating is more and more frequently employed,[66] which also increases the pultrusion rate. The most important characteristics of any heating system are uniformity of the temperature field distribution and its regulation (see Fig. 8.32).

Figure 8.32 Forming die for round bodies.

8.3.11 Pulling devices

Pulling devices have developed from the simplest forward thrust exercised via a wire rope or chain wound on a drum. At present, primarily two types of pulling devices are used.

A mechanism like a caterpillar track employs two caterpillar bands that rotating in opposite directions, with pads fixed on them that clamp and pull the cured profile. This method is preferable thanks to the large contact area with the profile being pulled, low cross and shear loads on the material, etc. Its disadvantage is that of using a large enough number of specially shaped clamping pads.

In simpler reciprocating mechanisms, only two pairs of pads are used for clamping each profiles type. One pad is pressed against the other by a hydraulic drive, and the profile is passed between them.[310] When the first pair of pads clamps and moves the profile to a given length, the other pair is running idle and returns to its initial position. For nonstop motion, idling occurs at a higher speed than the working cycle, and the second pair of pads clamps the profile before the first pair unclamps it (see Fig. 8.33).

8.4 General methodology of Setting Molding Regimes of Rod Elements

Selection of rational molding regimes involves a number of factors. During the first stage, the main structural and geometrical characteristics of the material are chosen according to its purpose and major restrictions imposed on the process are taken into account.

Figure 8.33 Reciprocating pulling device.

Under certain temperature conditions (see Chap. 3), the needed degree of solidification of the article is reached in the die and upon additional thermal treatment. The structure of the materials must comply with the set physical and mechanical characteristics attained under minimal pulling force.

The algorithm of choosing process regimes must include the following stages:

1. The main structural characteristics of the material are calculated, including thickness of its layers, distribution of reinforcing materials between the ingredients in the cross-section, filling degree, and required reinforcement angle of the winding layer.
2. Force parameters of the process, such as tension force of the transversal and winding reinforcement are defined on the base of recommendations put forward in Chaps. 5, 6, and 7.
3. Thermal treatment regimes are formulated according to investigation results set forth in Chap. 3.
4. Kinematic parameters of molding are estimated, including critical pulling rate, correlation of pulling, braiding and winding velocities.
5. The total pulling force is determined in accordance with the method presented in Chaps. 5 and 6.
6. At the final stage, it is estimated whether it is possible to mold the developed profile with account of restrictions formulated in Chap. 8. Respective design solutions are taken on the choice of lodgments of the pulling device, structural parameters of the material are adjusted, and kinematic characteristics of the device.

8.5 Conclusions

1. Proceeding from the analysis of the braiding–pultrusion process of complex-profile composite products, it should be noted that all major operations of the pultrusion process intended for a unidirectional composite are included and involve additional operations of molding the braiding layer, preforming, and stitching of the semifinished product.
2. The chief process regimes have been determined, and a calculation method for its parameters has been developed.
3. A principal scheme has been designed on the basis of the mentioned sequence of operations for the technological aspects of the concerned process.
4. Recommendations on the calculation and choice of special devices and equipment for the technological process, realizing the set characteristics and design parameters, are presented.
5. Based on the analysis of the stress–strain state of the rod under loading, the limiting clamping forces inflicting minimal damage to the material have been determined under the required pulling force. A design solution of the corresponding grips is proposed.

Chapter

9

Investigation of Physicomechanical Properties of Composites and Structures on Their Base

Physicomechanical properties of materials are affected significantly by the high velocity of the braiding–pultrusion process. Pultrusion alone and its effects on materials produced have been studied in numerous works.[137,202] Separate questions of structural optimization[88,214,215,385] and its stress–strain state have also received a great deal of attention.[45,209,271] The dynamic properties of pultruded products[48,49,199] and their characteristics under static loading[41,90,101,178,267,362] have been analyzed. A series of papers has been devoted to investigations of different types of beam structures.[30,34,42,44,89,90,163,199,263,272,299,300,340,363,364] Problems of operation of pultruded structures and materials under prolonged loading have been considered at length in.[11,27,314]

Because rod elements experience various loads, we consider their mechanical properties under the effects of different factors, such as pulling rate, temperature in the forming die, additional treatment, correlation of layers within the material structure, and reinforcement angles. Solid rods of various diameters as well as tubular rods and complex-profile double-T parts were used in the experiments.

9.1 Preparation of Experimental Samples

Samples with different structural parameters were fabricated using a braiding–pultrusion machine. The reinforcement scheme of both solid and tubular rods consists of longitudinal, transverse, and braided layers.

The structure of double-T parts has been estimated taking into account the additional stitching procedure. The reinforcement degree of each layer was varied under the preserved geometrical parameters of the articles and the preset force regimes considered earlier.

Test samples were cut from the rods directly at the outlet of the forming die upon additional treatment prior to feeding to the pulling device; other samples were made of rods that had completed thermal treatment. The samples were tested using standard testing machines. Recommendations expounded in[3,18,19,160,165,328,344,345,347] were followed during testing. Deformations were recorded by strain gauges and indicating gages. The results were processed in accordance with known methods of statistical analysis.

9.2 Effect of Process Parameters on Material Strength at Interlayer Shear

One of the main characteristics of composite materials is their resistance to interlayer shear. As indicated in Chap. 3, to raise efficiency and cut energy expenses, the binder is not cured completely at the outlet but is cured only to a glass transition state, at which its deformation and strength properties are at a minimal level that can preserve the article shape and eliminate damages at final drawing. It is emphasized that the pulling mechanisms clamp the article with considerable forces and that the material will still be thermally treated before it enters the pulling mechanism. However, this stage does not negate further final thermal treatment of the article after complete shaping and cutting. The article is then treated in a separate heating aggregate to finally set the binder. Heating and cooling of the article during additional thermal treatment lead to post-curing and physical annealing of the polymer matrix and its structural stabilization.[310,430] As a result, residual stresses in the material become lower and the glass transition temperature of the matrix rises, thus elevating its stress and strength properties, thermal resistance, and, consequently, the characteristics of the article as a whole.

It is expedient to estimate the interlayer shear using solid unidirectional rods with a rectangular cross-section. The values obtained can be further used to evaluate characteristics of composite materials, the effect of the types of their components, interactions at the interface, molding, and thermal treatment regimes.

The process parameters for fabrication of solid profiles were based on the results presented in previous chapters. The aim was to reach the required continuity of the material, surface finish, and filling degree. Because the interlayer shear resistance is highly sensitive to matrix

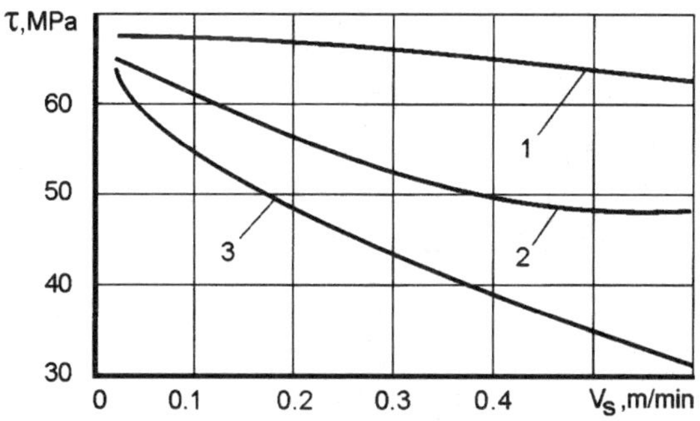

Figure 9.1 Effect of drawing rate on breaking stress at interlayer shear of glass fiber-reinforced plastics under $T = 240°C$: (1) after final thermal treatment at $T = 200°C$, $t = 3$ h; (2) after preliminary thermal treatment at $T = 200°C$, $t = 20 - 30$ min; (3) upon leaving the forming die.

properties, it can be an indicator of process quality at different stages. For investigations, samples 8 mm in diameter based on glass fibers and epoxy binder were used. Interlayer shear resistance was measured in the samples subjected to three sequential stages of thermal treatment, namely, immediately after leaving the forming die, at pretreatment, and at final thermal treatment following the usual technological regimes. Test results are presented in Fig. 9.1.

As seen in the figure, the strength of the material is impaired with increasing pulling velocity through the die because of lowered solidification degree of the matrix. Intermediate thermal treatment helps, in some cases, to improve interlayer strength, which appears sufficient for qualitative clamping in the pulling mechanism. It is important to make the right choice of the pulling velocity; on the one hand, increased pulling velocity essentially reduces the pulling force (i.e., energy consumption is diminished) while, on the other hand, impaired mechanical characteristics may result in impossibility of clamping the article in the pulling mechanism. After final thermal treatment, the strength of the material reaches a constant value (curve 1).

The mentioned data are in agreement with microstructural analyses. In Figs. 9.2 and 9.3, one can see microstructure of the rod material pultruded at a velocity of 0.8 m/min. This structure shows a considerable amount of microdefects at the interface. As the velocity falls to 0.4 m/min, the quality of the material improves considerably (Fig. 9.4), which is confirmed by reduced amount of defects.

The material was preliminarily treated in a special aggregate during the time the article is pulled after leaving the die and before entering the

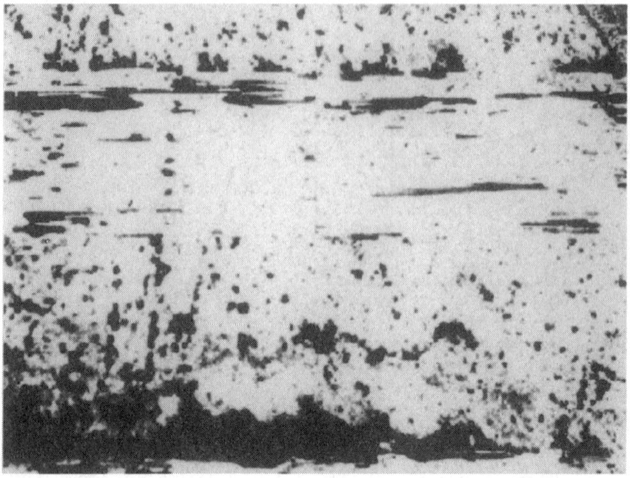

Figure 9.2 Microstructure of the rod material at pultrusion velocity 0.8 m/min.

pulling mechanism. The interlayer shear dependence on pretreatment temperature is shown in Fig. 9.5. As is seen, an optimum temperature can be chosen for a given velocity of the article to attain the best strength indices.

Based on experimental data, it is possible to define the optimum relation of the article velocity to pretreatment temperature and optimum final physicomechanical properties.

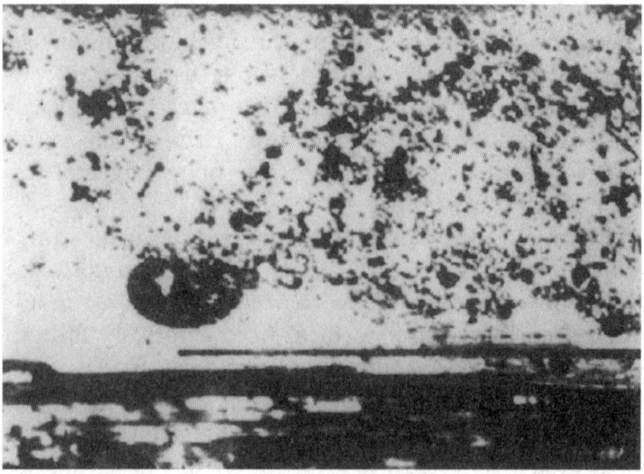

Figure 9.3 Structural defects in the material.

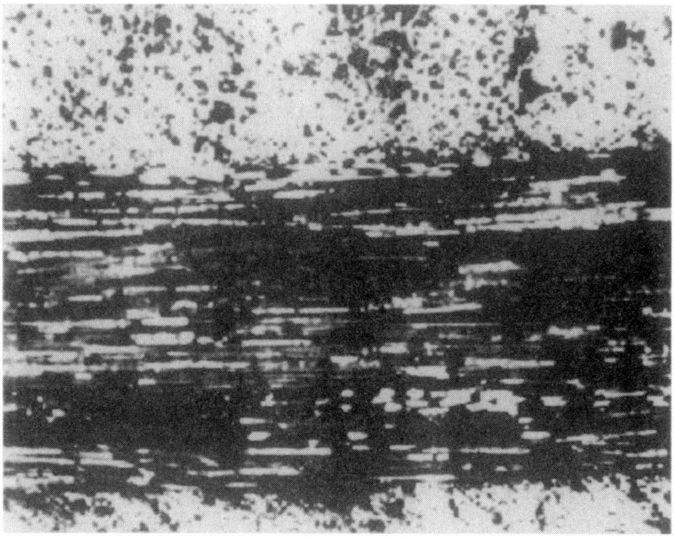

Figure 9.4 Microstructure of the rod material at pultrusion velocity 0.4 m/min.

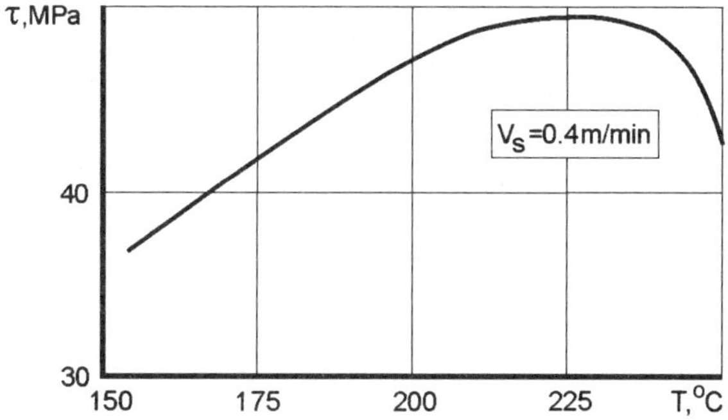

Figure 9.5 Effect of pretreatment temperature on interlayer shear.

9.3 Investigation of Composite Properties at Axial Compression

As indicated above, the materials fabricated by the braiding–pultrusion method are most effectively used in rod-like structures operating under severe tensile and compressive loads. In this connection, the relation of the composite properties to their structural parameters should be studied for each individual case. Most explicit is in our view the interrelation of structural and process characteristics of the articles at axial compression. Stiffening of the outer surface of the rod by a braided reinforced layer turns out to be a means of regulating the properties of the entire article. This outer layer may impair the strength characteristics of the article in axial direction. In contrast, shear and transversal strength, as well as stiffness resistance to longitudinal delamination, are improved. So, it is reasonable to determine the optimum correlation of properties of the stiffening and longitudinal layers, estimate the transversal stresses and strains evoking delamination, and examine breakage of the material as a whole.

Problems of rational stiffening of rod structures have been studied by a number of researchers.[3,45,73,78,100,137,287,320,321,412,419] For instance, a relation derived in Ref. 346 on the base of the energy theory for critical stresses leading to separation of the longitudinal reinforcement from the circular one at compression looks like

$$\sigma_C = 0.916\,E^*\left[\left(\frac{h}{r}\right)^2 + \aleph\left(\frac{h}{r}\right)^{-1}\right]^{1/2} \tag{9.1}$$

where $\aleph = 12\gamma/E^*r$
E^* = efficient circular elasticity modulus
h = wall thickness of the tubular rod
r = inner radius of the tubular rod
γ = specific delamination energy defined in tests on the separation of composite strips

For a compressible, solid, annular composite rod, the delamination stresses can be determined[141] as follows:

$$\sigma_d = E^*K\left(\frac{h}{L}\right)^2 \times \left(1 + \sqrt{1 + \zeta\left(\frac{h}{L}\right)^{-5}}\right) \tag{9.2}$$

where $\zeta = 12\gamma/E^*L\,(1-\nu)K$
K = factor dependent on spall shape
L = delamination length
ν = Poisson's ratio

The critical stress of a rod covered by an isotropic material at compression is considered in Ref. 346 based on the energy theory

$$\sigma_C = \frac{1}{3}E_Z\pi^2\eta^2\left(1 + \frac{G_{Z\Theta}}{\pi}\sqrt{\frac{\gamma}{E_Z L}}\,\eta^{-5}\right)$$
$$L = 0.815\,\pi\left(\frac{t_b^5 E_Z}{\gamma}\right)^{1/4} \qquad \eta = \frac{t_b}{L} \tag{9.3}$$

where L and t_b are the length and thickness of isolated isotropic strip, respectively, and E_Z and $G_{Z\theta}$ are the elasticity moduli.

It has been established that, as a result of compression of tubular rods of circular and square sections, there arises a mechanism of stability loss, i.e., lengthwise delamination of the Chinese lantern type. To exclude this effect, it was proposed in Refs. 154 and 346 that thin, reinforced circular layers be introduced with reinforcement angles above $\pm 10^\circ$. This ameliorates the shear modulus $G_{Z\theta}$, and the specific delamination energy γ is increased.

A calculation method of the stress–strain state of a hybrid composite rod with stiffening layers experiencing compressive loading was developed.[103,115,116] The calculations have proved that a combination of hybridization with stiffening reduces cross-strains and moves aside the delamination threshold.

Analysis of the literature shows that introducing additional stiffening layers in the structure of compressed rods is strongly recommended, although it does impose certain limitations on the process. It is advisable but, in fact, improbable to introduce only a transversal reinforced layer at an insignificant reinforcement angle because the presence of the outer braided layer is necessary for the pultrusion process.

To analyze the relation of the bearing longitudinal and stiffening layers, let us consider a rod compressed by a longitudinal force. It is presumed that a unidirectional rod compressed by a force P_0 in the reinforcement direction and the compressive strains σ_Z^- that appear in this case result in its lengthwise deformation, the value of which is calculated according to Hook's law

$$\varepsilon_Z = \frac{\sigma_Z^-}{E_Z} = \frac{P_0}{FE_Z} \tag{9.4}$$

where σ_Z^- is the breaking compressive strength, E_Z is the elasticity modulus in the reinforcement direction, and F is the cross-sectional area.

The relative deformation in the radial direction is calculated by the relation

$$\varepsilon_r = \nu_{r\theta}\varepsilon_Z \tag{9.5}$$

When a unidirectional rod is compressed by a value ΔL with force P_0, the following work is performed:

$$A_0 = \frac{P_0 \Delta l}{2} \tag{9.6}$$

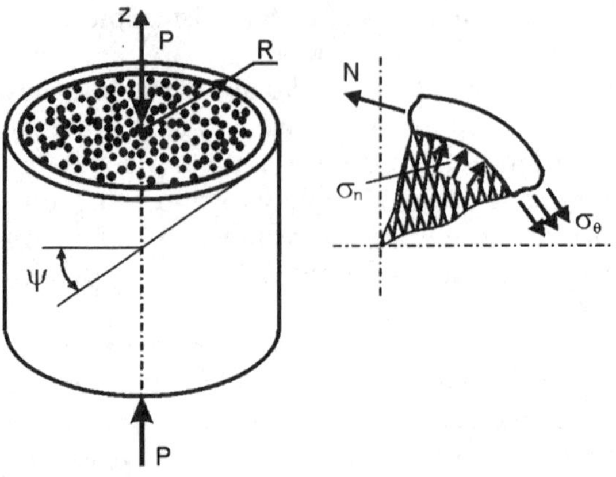

Figure 9.6 Diagram showing the variables used in the calculation of stresses in the stiffening layer.

Let us now take a rod of a similar cross-section and outer stiffening layer (Fig. 9.6). To compress it by value ΔL, one should apply the force

$$P = P_0 + \Delta P \tag{9.7}$$

and perform the work

$$A = A_0 + \Delta A \tag{9.8}$$

where $\Delta A = \Delta P \Delta L/2$, which is the work spent on deformation of the stiffening layer.

In presumption of negligibility of the braided layer size and the symmetry of the reinforcement of this layer relative to the article's long axis, we analyze the dependence of effective stresses on characteristics of a unidirectional transversal layer that stiffens the longitudinal reinforcement. The deformation of the stiffening layer will then be

$$\Delta L = -\frac{\nu_{r\theta} P_0 L}{E_{r\theta}} \tag{9.9}$$

where $L = t_0/\sin\psi$, t_0 is the pitch of laying the stiffening layer, and ψ is the winding angle.

As far as the materials inside and outside the rod deform jointly, a stretching force N arises in the reinforcement of the stiffening layer which is interrelated with deformation of this layer ΔL through the equation

$$\Delta L = \frac{NL}{E_N F_N} \tag{9.10}$$

where E_N is the longitudinal elasticity modulus of the stiffening layer along axes of its elasticity symmetry and F_N is the cross-sectional area.

The potential energy is accumulated in the winding reinforcement at its deformation

$$A_H = \frac{N^2 L}{2E_N F_N} = \Delta A \tag{9.11}$$

The stretching force N induced in the winding reinforcement by the compressive force is found from Eqs. (9.9) and (9.10)

$$N = -2P\frac{\nu_{\mathrm{rz}} E_N F_N}{2E_{z\theta}F + \nu_{\mathrm{rz}}^2 \frac{\delta t_0 E_N}{\cos\Psi}} \tag{9.12}$$

where δ is the stiffening layer thickness.

Tensile stresses in the winding reinforcement will be expressed through

$$\sigma_\theta = -2\sigma_Z^- \frac{E_N}{E_Z} \times \frac{\nu_{\mathrm{r}\theta}\nu_{\mathrm{rz}}}{\left(2 + \upsilon_{\mathrm{r}\theta}^2 \frac{E_N}{E_Z} \times \frac{\delta t_0}{F\cos\Psi} \times \frac{\nu_{\mathrm{rz}}}{\nu_{z\mathrm{r}}}\right)} \tag{9.13}$$

Normal stresses σ_{n} appearing at the interface with the stiffening layer are dependent on its rigidity, elasticity and deformation properties. Bearing in mind Eqs. (9.11) and (9.13) we can calculate them as follows:

$$\sigma_{\mathrm{n}} = 1.81\sigma_Z^- \frac{\delta}{R} \times \frac{E_N}{E_Z} \times \frac{\nu_{\mathrm{r}\theta}\nu_{\mathrm{rz}}}{\nu_{z\mathrm{r}}\left(2 + \nu_{\mathrm{r}\theta}^2 \frac{E_N}{E_Z} \times \frac{\delta t_0}{F\cos\Psi} \times \frac{\nu_{\mathrm{rz}}}{\nu_{z\mathrm{r}}}\right)} \tag{9.14}$$

Proceeding from the above, compression of a unidirectional profile stiffened by an outer winding layer generates normal stresses at the interface between the layers and tensile stresses in the outer layer. As it follows from Eqs. (9.13) and (9.14), these stresses depend on the stiffening layer thickness and characteristics of the materials used. In Figs. 9.7 and 9.8, the dependencies of the stresses on the stiffening layer thickness and article size are presented.

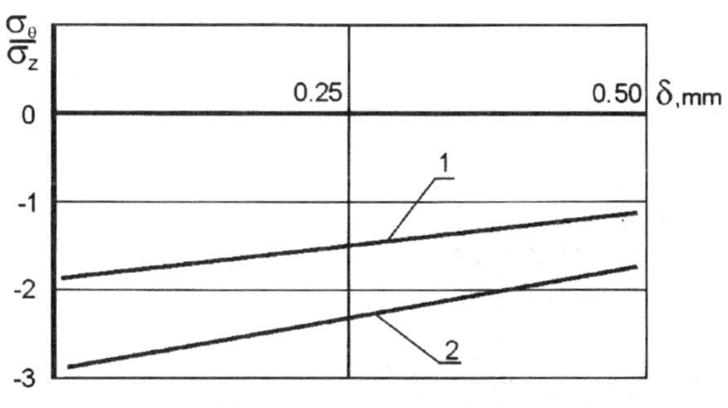

Figure 9.7 Effect of the winding layer parameters on tensile stresses in the stiffening layer: (1) glass-fiber winding; (2) organic fiber winding.

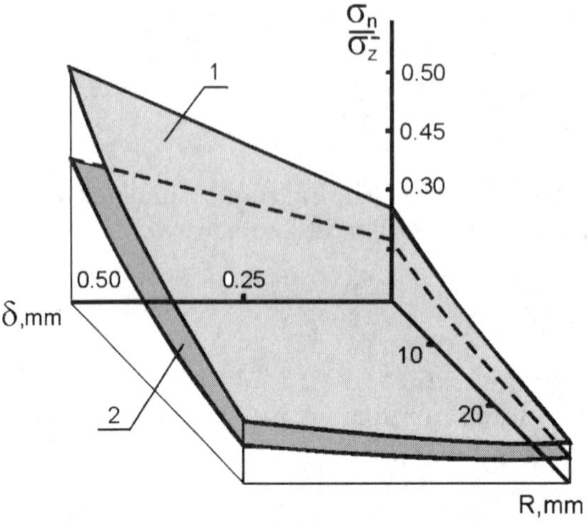

Figure 9.8 Effect of the winding layer parameters on normal stresses at the interface with the stiffening layer: (1) glass-fiber winding; (2) organic fiber winding.

In Figs. 9.9 and 9.10, calculation results are presented of these stresses by relations (9.13) and (9.14) according to Ref. 116 for a glass plastic rod with a layer of organic plastic.

As it follows from above data, radial stresses are not high in contrast to limiting ones, but their effect on relative cross strains is essential. Besides, as indicated in Ref. 154 they cause the longitudinal cracks that initiate failure of the strips, resulting in diminished stability. Note that for the solid rods, the probability of delamination is equal in the radial and circumferential directions. It also follows from Fig. 9.10 that there exists some critical value of the stiffening layer packing $\psi_C = 27°$, at which stresses in the reinforcement direction of the stiffening layer

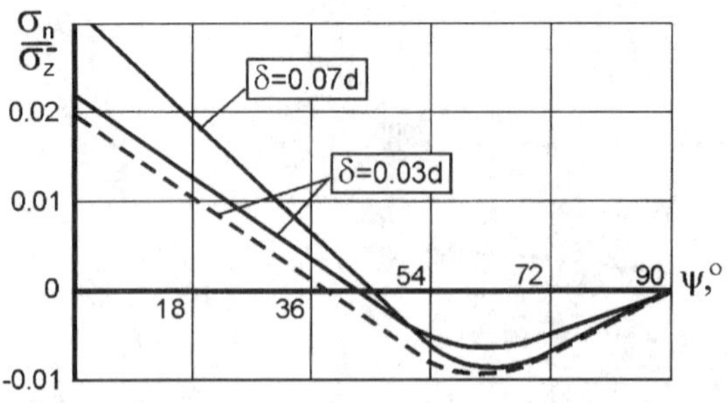

Figure 9.9 Effect of reinforcement angle of the stiffening layer on radial stresses at compression: (—) according to[116]; (- - -) by the relation (9.14).

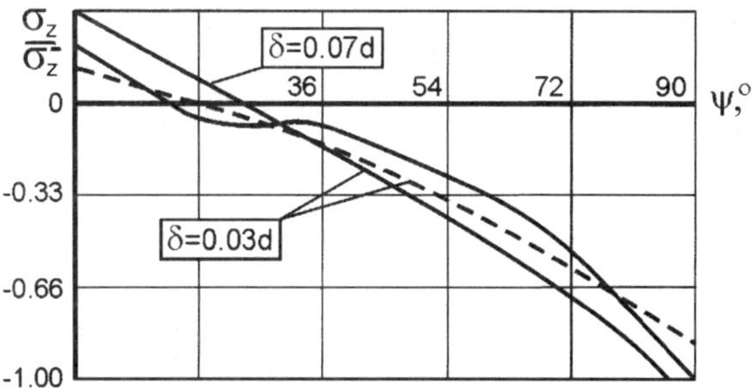

Figure 9.10 Dependence of stresses in the stiffening layer on its reinforcement angle: (—) according to Ref. 116; (- - -) by the relation (9.13).

change their sign. When $\Psi > \Psi_C$, cross-deformation inside the rod is higher than in a unidirectional one. This shows that the layer introduced in it at angle $\Psi > \Psi_C$ is not stiffening and even promotes delamination of the rod. In this case, longitudinal compression of the rod induces tensile stresses in the stiffening layer dependent on the elasticity modulus of this layer and its thickness.

Hence, failure of the material may be initiated by one of the following reasons:[287,393] (i) cross-strains of the main material become limiting; (ii) tensile stresses in the layer reach breaking strength at tension; or (iii) stresses in the longitudinal reinforcement reach breaking stress at compression σ_{zn}. Thus, we can find breaking stress of the rod covered by an outer stiffening layer

$$\sigma_{SZ} = \sigma_{SZ}^{(0)} \min \left\{ \frac{\varepsilon_{\lim}}{\varepsilon_1}; \frac{\sigma_{SL}^{(0)}}{\sigma_0}; \frac{\sigma_{zn} F}{\sigma_{SZ}^{0} F_L} \right\} \tag{9.18}$$

where $\sigma_{SZ}^{(0)}$ = breaking strength of a unidirectional rod at compression
ε_1 = cross-strain of the main material of the rod at stress $\sigma_{SZ}^{(0)}$
$\varepsilon_{\lim}$ = limiting value of ε_1
$\sigma_{SL}^{(0)}$ = breaking strength at tension
σ_0 = stress in the layer under stress $\sigma_{SZ}^{(0)}$
σ_{zn} = breaking strength of longitudinal reinforcement at compression
F = area of the rod
F_L = area of the longitudinal reinforcement

The theoretical results were compared to experimental data for 50-mm samples from various parts of the rods produced by a braiding–pultrusion machine. The load was applied at deformation rate of

TABLE 9.1 Effect of the Method of Terminating Sample Faces on Compressive Strength of the Rods

	Unidirectional			With a stiffening layer		
Type of termination	$\sigma_{\bar{Z}}$, MPa	S, MPa	ω, %	$\sigma_{\bar{Z}}$, MPa	S, MPa	ω, %
Simply supported or polished	232	31	13.4	201	25	12.5
Inserted in metal cups	295	20	6.9	249	23	9.2
Filled up in metal cups	382	6	1.6	389	12	6.2
Filled up in metal cups and moving under loading over guides	396	17	4.3	416	22	5.4

**S* is the mean square deviation of test results.

0.2–1.0 mm/min. First, the samples were loaded at a force corresponding to 15–20 percent of anticipated short-term static strength of the material, after which the load was lowered down to 0.05 of the breaking force. This condition was taken as the initial state.

Unidirectional rods and the rods with a stiffening braided layer were used in testing. To achieve values closest to the actual strength of the materials at compression, different types of face clamping of compressed samples were studied. Test results are presented in Table 9.1. The results prove that the method of terminating faces of the rods effects perceptibly their strength values. It is seen that the samples with a stiffening layer are more sensitive to this factor than the unidirectional rods. Further investigations have shown that the material of these samples does not exhaust its bearing capacity but breaks as a result of loss of stability of the sample itself. To avoid this drawback, a tubular guide was mounted on the apparatus, over which metal cups could move and pour compound onto the faces of the rod. These cups preserve the relative coaxiality and assist in achieving more reliable strength characteristics.

To study the effect of the stiffening layer material and its thickness on the strength and delamination threshold of rod articles, 20 mm in diameter glass-reinforced plastic rods with a thick (0.1 to 0.6) mm outer stiffening layer of organic plastics were tested. The experimental results and theoretical data are presented in Table 9.2.

The results have shown that the braiding layer of 0.1 to 0.2 mm thickness fails as it reaches the breaking tensile strength. Note that the theoretical results showed good agreement in this case with experimental investigations.

As thickness of the stiffening layer increased above 0.3 mm, the longitudinal reinforcement lost its continuity. The breaking strength was in this case a little higher than the rated one. This is because the braided

TABLE 9.2 Effect of Stiffening Layer Thickness on Strength of Glass-Reinforced Plastic Samples at Compression*

Stiffening layer material	δ, mm	Theory		Experiment		
		σ_Z^-,MPa	σ_Z^-, MPa	S, MPa	Ω,%	Fracture type
Without a layer	—	—	396	17	4.3	Delamination
Organic plastic	0.1	409	428	21	5.1	Braiding rupture
	0.2	422	447	13	3.0	Braiding rupture
	0.4	458	483	34	7.5	Reduction of the main material
	0.6	472	496	17	3.7	Reduction of the main material

*Reinforcement angle is 45°.

layer hinders spalled parts of the material from losing stability. They lose their bearing capacity only after breakage of the outer layer followed by failure of the delaminated longitudinal main material.

A thickness increment of 0.6 mm in the layer did not result in the desired improvement of compressive strength as it led to contraction of the longitudinal reinforcement area and, consequently, to intensified stresses. The samples break in the plane where the maximum tangential stresses were acting at an angle 49° to the rod axis.

Proceeding from the above, it follows that introduction of a stiffening braided layer in the rods with continuous sections produced by the braiding–pultrusion method changes their stress–strain state and fracture behavior at compression. The efficiency of this layer depends upon both geometrical parameters and the relation of elastic constants of the materials as well.

9.4 Stability of Pultruded Compressible Tubular Rods

The bearing capacity of tubular rods under compressive loads has been estimated both theoretically and experimentally. With this aim, the effect of structural and process parameters of the articles on the critical load at the moment of losing stability have been studied. To keep to the method described in Ref. 381 the critical load can be calculated as follows:

$$P_C = \frac{\pi^2 m^2 R^2 B}{2L_r B_{22}\left(1 + \frac{\pi^2 m^2 R^2 B}{L_r^2 B_{22} B_{33}}\right)} \tag{9.19}$$

Here the parameter m defines the conditions of rod fastening. In case of pivot support, $m = 1$; at immovable termination of the faces, $m = 2$.

Thus, the generalized membrane rigidities with account of the material cleavage will be

$$\begin{aligned} B_{11} &= \sum_{i=1}^{4} Q_{11}^{(i)} H_2^{(i)} h_i \\ B_{22} &= \sum_{i=1}^{4} Q_{22}^{(i)} H_1^{(i)} h_i \\ B_{12} &= \sum_{i=1}^{4} Q_{12}^{(i)} h_i \\ B_{33} &= \sum_{i=1}^{4} Q_{33}^{(i)} h_i \\ B &= B_{11} B_{22} - B_{12} \end{aligned} \tag{9.20}$$

and values $H_1^{(i)}$ and $H_2^{(i)}$ will be

$$H_{1,2}^{(i)} = \frac{\left(1 + \frac{Z_i - Z_{i-1}}{2R_{1,2}}\right)}{\left(1 + \frac{Z_i - Z_{i-1}}{2R_{2,1}}\right)}$$

where Z_i = distance from coordinate origin on the lower boundary of the laminated material to the next layer

$R_{1,2}$ = inner and outer radii of the rod

L_r = rod length

R = radius of the middle surface

The components of the rigidity matrix are found using properties of the initial ingredients and orientation of the layer toward the article axis.

For the investigations, rod faces were simply supported via pouring a cold-setting compound in special caps with ball shanks. Tubular rods of 20 mm inner diameter and 1.5 mm wall thickness were manufactured. The varied parameters were the transversal-to-longitudinal layer ratio, the total filling degree of the article, and the braiding angle of the outer layer. Along with other properties, the drawing rate and thermal treatment conditions were estimated.

The dependencies of the critical load on the outer layer reinforcement angle in the moment the tubular rod looses its stability P_C are presented in Fig. 9.11. With increasing reinforcement angle, the critical load decreases according to both theoretical and experimental results. This circumstance is explained by diminishing reduced elasticity modulus of the material with increasing reinforcement angle.

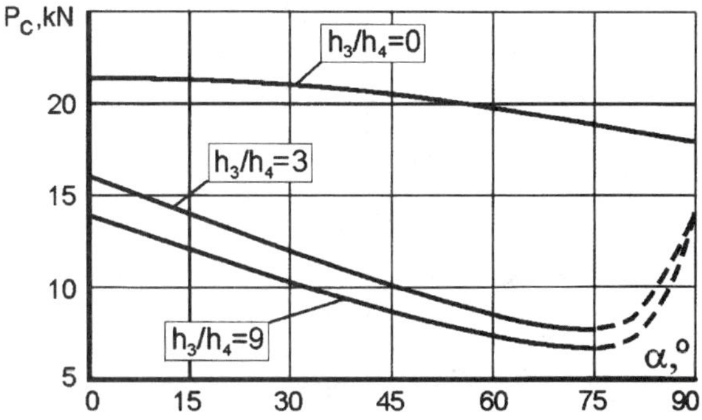

Figure 9.11 Effect of reinforcement angle of the braided layer on critical load at stability loss of tubular rods.

The filling degree affects perceptibly P_C value as well. As it follows from the theoretical data presented in Fig. 9.12, with increasing filling degree P_C augments, too. Experimental data are not in full agreement with theoretical results: P_C goes down already at filling degree values surpassing 0.5–0.55. Probably, various structural defects, such as unglued places, air inclusions, etc., are the reason of local loss of stability.

The effect of the longitudinal-to-transversal layer ratio on the stability of tubular rods is illustrated in Fig. 9.13. A general tendency in diminishing critical load with reducing longitudinal layer share is exhibited within the total invariable cross-sectional area occupied by these layers independently of characteristics of the braided layer.

Highly indicative is the dependence of the critical load on the article drawing rate (Fig. 9.14). In principle, it would be justified to speak about optimum drawing rate value $V_S > 0.3$ m/min, independently of

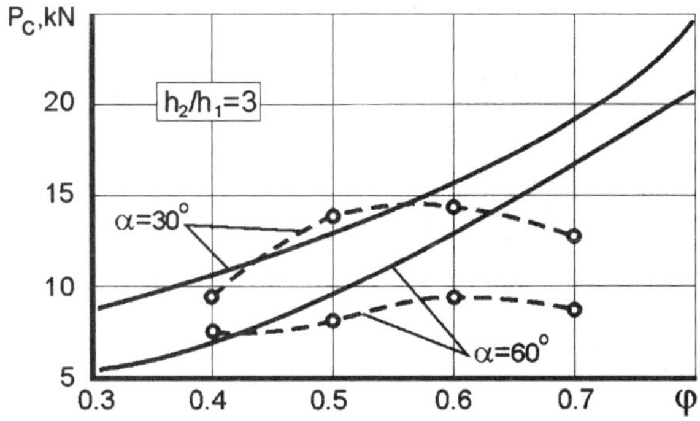

Figure 9.12 Theoretical and experimental dependencies of the critical load at tubular rod collapse on filling degree and reinforcement angle of the outer layer: (–) theory; (- - -) experiment.

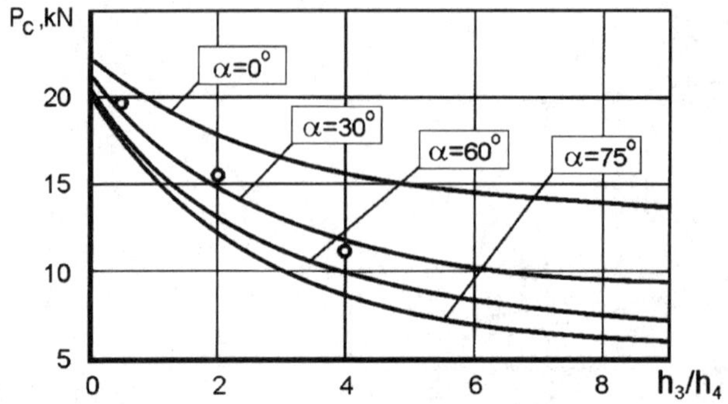

Figure 9.13 Effect of longitudinal to transversal layer ratio on the critical load tubular rod collapse: (○) experimental values of P_C at $\alpha = 30°$.

the type of thermal treatment. This kind of dependence is true under low pulling velocities when separate fibers start to break in the material structure leading to propagation of local failure that changes much the stress-strain state, which may be a reason of further breakage.

Loss of stability of tubular rods manufactured by the braiding–pultrusion method differs from the unidirectional ones. The latter fail as they lose stability because of lengthwise delamination of the reinforcement. As for the tubular rods, with an outer braided layer, they return into their initial state without any residual stresses as soon as the load is removed.

Analogous results obtained during investigations of 25.4 mm in diameter tubular rods with wall thickness of 1.6 mm are illustrated in Figs. 9.15 and 9.16.

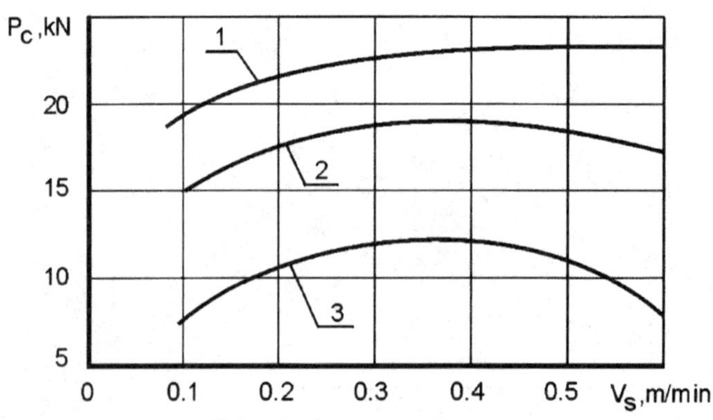

Figure 9.14 Effect of drawing rate and thermal treatment regimes on critical load at tubular rod collapse: (1) after final thermal treatment; (2) after pretreatment in the die; (3) upon leaving the die.

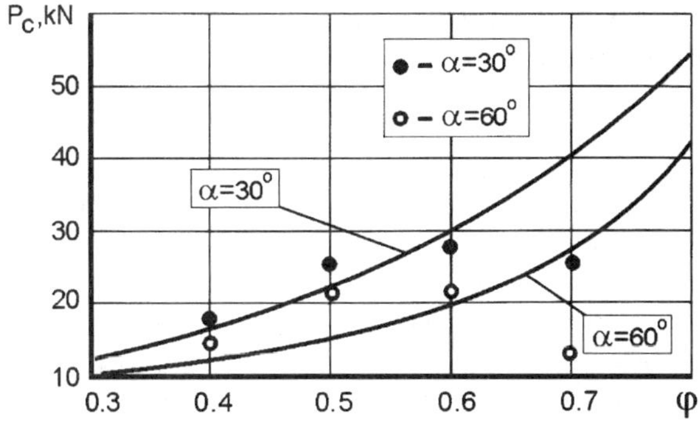

Figure 9.15 Effect of filling degree on critical load of tubular rod (Ø25.4) mm as it loses stability.

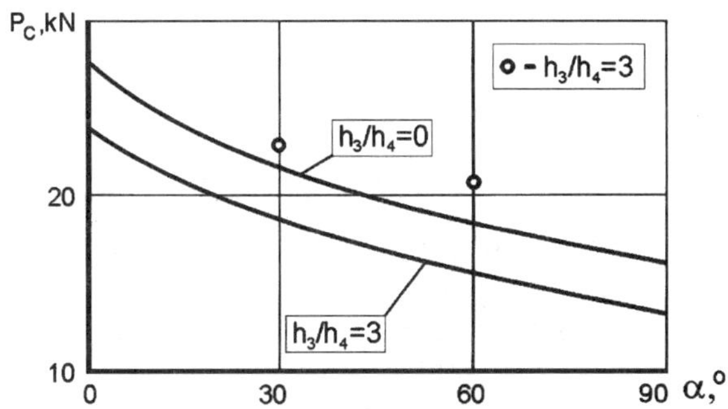

Figure 9.16 Effect of reinforcement angle of the braided layer on variations of the critical load in Ø25.4 mm rod.

9.5 Bearing Capacity of Pipes Produced by Braiding–Pultrusion under Inner Pressure Effect

To estimate bearing capacity of the pipes, let us take a cylindrical shell loaded with an inner pressure and tensile force whose value is conditioned by the effective pressure and also depends on the open flow area of the pipe. The stresses acting in this case are calculated by the following relations:[381]

$$\sigma_1 = \frac{E_x(1+\nu_{X\theta})(P+q)\,R_\mathrm{p}}{2\sum\limits_{i=1}^{K} h_i\left[E_x^i\left(1+\nu_{x\theta}^i\right)+E_\theta^i\left(1+\nu_{\theta x}^i\right)\right]}$$
$$\sigma_2 = \frac{E_0(1+\nu_{\theta x})(P+q)R_\mathrm{p}}{2\sum\limits_{i=1}^{K} h_i\left[E_x^i\left(1+\nu_{x\theta}^i\right)+E_\theta^i\left(1+\nu_{\theta x}^i\right)\right]} \quad (9.21)$$

where P and q = force at pitch and inner pressure
E_x, E_θ, and $\nu_{\theta x}$ = elastic constants of the materials
K = number of layers
E^i and ν^i = elastic constants of the ith layer
R_p = inner diameter of the pipe
h_i = thickness of the ith layer
h = thickness of the whole packet

In addition, according to Ref. 325 we can estimate the strength of each layer of the shell by presuming that the inner longitudinal layer starts to break together as the binder crack and, consequently, is the reason for losing tightness. Hence, the strength condition will be of the form

$$\sigma_1^2 d_1 + 2\sigma_1\sigma_2 d_2 + \sigma_2^2 d_3 = \frac{\Delta^2}{h^2} \tag{9.22}$$

where

$$d_1 = (1-\nu_\mathrm{C}^2)\sigma_\mathrm{r}^{-2}\frac{a_2^2}{(R_\mathrm{C}^+)^2} + \frac{a_3^2}{T_\mathrm{C}^2}$$

$$\Delta = Q_{11}Q_{22}Q_{66} - Q_{12}^2 Q_{66}$$

$$d_2 = (1-\nu_\mathrm{C}^2)\sigma_\mathrm{r}^{-2}\frac{a_2 b_2}{(R_\mathrm{C}^+)^2} + \frac{a_3 b_3}{T_\mathrm{C}^2}$$

$$d_3 = (1-\nu_\mathrm{C}^2)\sigma_\mathrm{r}^{-2}\frac{b_2^2}{(R_\mathrm{C}^+)^2} + \frac{b_3^2}{T_\mathrm{C}^2}$$

The factors a_2, b_2, a_3, and b_3 and the stress concentration factor σ_r^- have been calculated elsewhere;[325] R_C^+ and T_C are the tensile and shear strength of the binder, respectively.

Once the binder fails in the first layer, its characteristics in the direction perpendicular to fibers are taken to be zero, after which the strength condition of the transversal and braiding layers can be written as

$$\begin{aligned}&\sigma_1(Q_{22}Q_{66}\cos^2\alpha - Q_{12}Q_{66}\sin^2\alpha)\\ &+\sigma_2(Q_{11}Q_{66}\sin^2\alpha - Q_{12}Q_{66}\cos^2\alpha) = \frac{\Delta}{h}\varepsilon_\mathrm{f}^+\end{aligned} \tag{9.23}$$

where α is the reinforcement angle of the layer; Q_{ij} are components of the rigidity matrix; and ε_f^+ is the ultimate strain of reinforcing fibers.

Thus, we can calculate the pressure value corresponding to failure onset, i.e., loss of tightness and total collapse of the pipe, by sequential substitution of stress values from Eq. (9.21) into strength conditions and taking into account the P-to-q ratio.

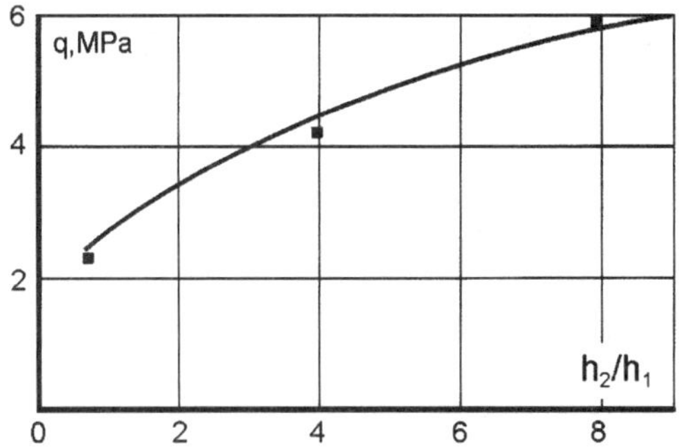

Figure 9.17 Breaking strain variation after final thermal treatment depending on transversal-to-longitudinal layer ratio.

In experimental studies, we employed a special hydraulic setup where pipe samples, which were plugged tightly from both sides, were loaded by inner pressure. The pressure under which the tightness of the pipe was violated and ultimate breaking pressure of the pipe were estimated. The pipe length was within $L_p > 10d$ to exclude edge effect. In Fig. 9.17, theoretical and experimental results of studying the breaking pressure dependence on the transversal-to-longitudinal layer ratio are presented. Every point on the curves is averaged from at least 5 replicate tests. It is seen in the diagram that pressure increases as the transversal reinforcement share increases. This forms the basis for predicting the behavior of such products under service loads.

With increasing reinforcement angle (the angle between the article axis and the reinforcement direction) at a constant ratio of the layers, the breaking pressure tends to grow (Fig. 9.18).

The samples for studying the effect of process parameters were cut from a continuous profile at the outlet from the die after their pretreatment. In a number of cases, such samples could not withstand the clamping force of the drawing mechanism due to very low transversal stiffness. The samples underwent two types of tests: sealing and resistance trials using a sealant. The dependence of pressure at which

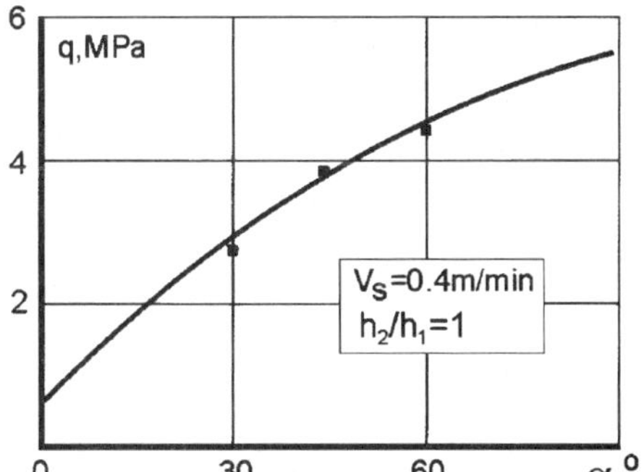

Figure 9.18 Reinforcement angle effect on critical pressure.

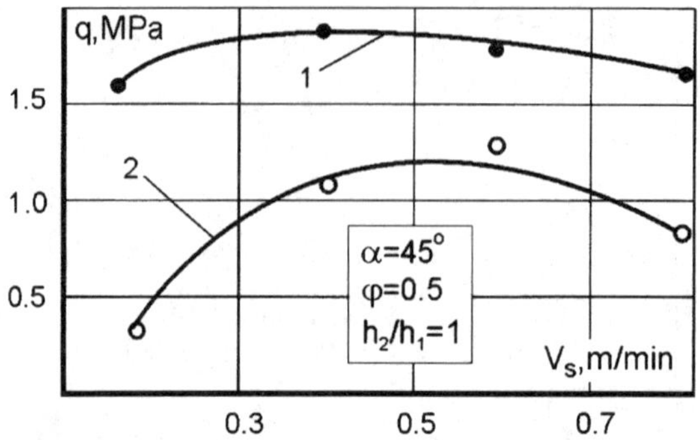

Figure 9.19 Drawing rate and thermal treatment effect on pressure corresponding to violation of tightness: (1) after additional thermal treatment at $T = 220°C$; (2) at the outlet from the forming die.

tightness of the pipe is lost versus the drawing rate through the forming die is presented in Fig. 9.19.

Analysis of the curves has shown that, under low drawing rates, the pressures are not high at any type of thermal treatment. This can be most likely attributed to propagation of cracks in the structure and partial rupture of longitudinal fibers due to significant tension forces arising in this case. With a considerable increase in the drawing rate, i.e., $V_S = 0.6 - 0.7$ m/min, the strength is impaired as well, mainly due to the occurrence of air and gas voids that are considered to be stress concentrators. These phenomena are most clearly exhibited in the samples not subjected to additional thermal treatment and, consequently, in those having a lower curing degree of the binder.

Results of investigations of tightness as a function of temperature in the forming die (Fig. 9.20) corroborate with conclusions that the optimum die temperature is 200–220°C. Note that a decrease in the temperature reduces the strength because of the diminished degree of

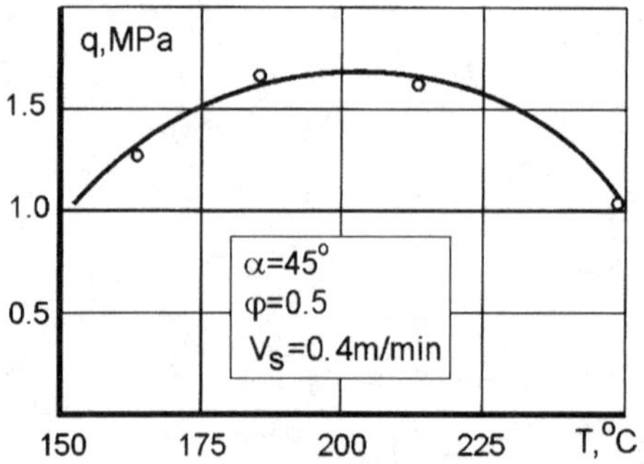

Figure 9.20 Pressure of tightness violation versus temperature of the forming die.

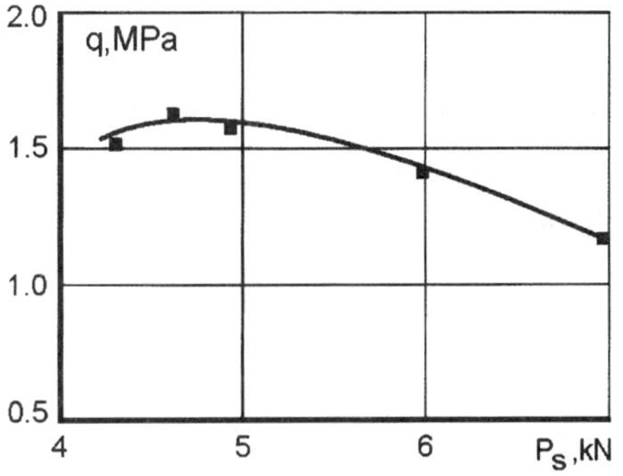

Figure 9.21 Pressure of tightness violation versus tension force of pretreated articles.

polymerization, while an increase in temperature impairs the strength as a result of the binder becoming more loose and its partial destruction.

Based on the data presented in Fig. 9.21, it is possible to estimate pressure variations at tightness violation as a function of tension forces. It is evident that with increasing tension force, the pressure drops, which is consistent with the results presented in Fig. 9.22; reduction of the pulling force leads to growing tension forces and, consequently, to worsened article characteristics.

The final thermal treatment in a separate heating aggregate does not alter in any way the critical pressure dependence on the drawing rate (Fig. 9.22). This fact makes possible the supposition that the reasons mentioned above to explain the variations in article properties are most feasible. Investigations of the preliminary thermal treatment temperature effect on bearing capacity of the pipes (Fig. 9.23) have proved that temperature does not influence the properties of the articles that are subjected to final thermal treatment.

The consistency of theoretical and experimental data cited in Table 9.3 was evaluated. It follows from the analysis that the proposed method assists in forecasting the behavior of articles under mentioned

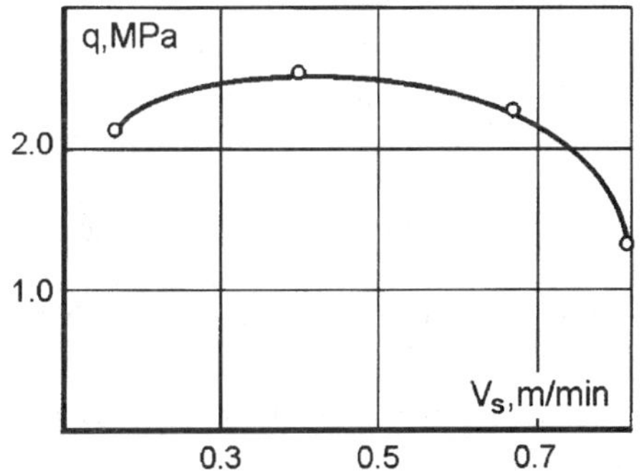

Figure 9.22 Effect of drawing rate on breaking pressure after final thermal treatment.

TABLE 9.3 Variations of Critical Pressure and Tightness Breaking Pressure as Functions of the Structural Parameters of 25.4-mm Diameter Pipes

		Tightness q, MPa			Strength q, MPa		
$\frac{h_1}{h_2}$	α°,	Theor.	Exp.	ω, %	Theor., q_P	Exp., q_P	ω, %
1.0	30	0.97	1.05	7.3	2.42	2.0	4.0
	45	1.12	1.03	11.0	3.23	2.7	5.5
	60	1.26	1.08	6.6	4.08	3.95	5.9
8.0	30	1.11	1.25	4.7	4.51	3.85	6.2
	45	1.32	1.35	8.2	5.86	6.0	7.1
	60	1.35	1.5	2.5	7.47	7.25	10.2

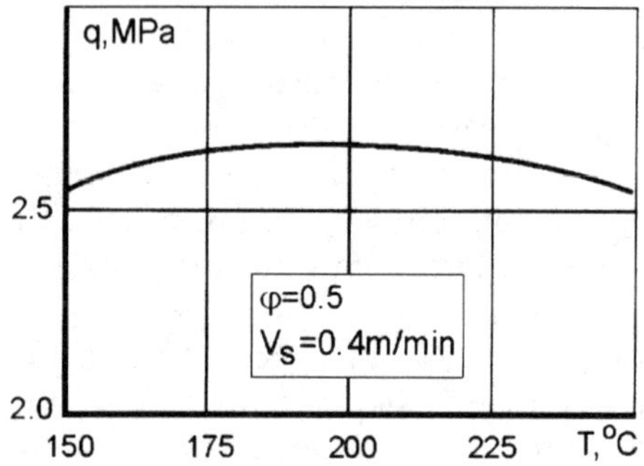

Figure 9.23 Effect of preheating temperature after final treatment on breaking pressure.

types of loading at a fairly high quality. The theoretical data deviate from those obtained experimentally by less than 19.5 percent.

9.6 Strength and Bearing Capacity Estimation of Complex-Profile Rod Articles

For a comparative analysis of complex-profile articles, experimental samples of double-T shapes have been fabricated of composite materials of unlike structural schemes realized during production. The samples have been subjected to service loads with the bending moment M_z = 1 kN m and lateral force $Q_y = 1$ kN. All samples are based on an epoxy carbon plastic. Their chief mechanical characteristics used as the basic design data are shown in Table 9.4.

The profile produced by the braiding–pultrusion method corresponds recommendations expounded in Sec. 4.3. The profile type, as well as its main structural and geometrical parameters are illustrated in Fig. 9.24 and Table 9.5

Aside from the profiles, for comparison we have fabricated unidirectional parts with a similar structure. They are usually produced on an apparatus based on a differential design principle, but they can be

TABLE 9.4 Mechanical Characteristics of Carbon Plastics

Properties	Characteristics of the packets	
	$\alpha = 0^\circ$	$\alpha = 45^\circ$
E_x, GPa	100	20
E_y, GPa	10	20
E_z, GPa	10	10
G_{xz}, GPa	4	4
G_{xy}, GPa	4	6
$\alpha_x \cdot 10^{-6}, 1/^\circ K$	0	3.46
$\alpha_y \cdot 10^{-6}, 1/^\circ K$	30	30
$\alpha_z \cdot 10^{-6}, 1/^\circ K$	30	3.46
μ_{xy}	0.35	0.664
μ_{yx}	0.035	0.664
μ_{xz}	0.35	0.35
σ_x^+, MPa	900	150
σ_x^-, MPa	700	150
σ_y^+, MPa	50	50
σ_y^-, MPa	120	150
σ_z^+, MPa	50	150
σ_z^-, MPa	120	150
τ_{xy}, MPa	30	30
τ_{xz}, MPa	30	50
τ_{zy}, MPa	30	30

fabricated by a continuous method as well. A part of the samples produced by the braiding–pultrusion method was stitched in the junction with the flanges by a Kevlar braid of linear density $T = 400$ tex., while the other part was fed into the forming die without stitching.

All profile articles were designed so as to fit the sizes of the profile cross-section and preserve the needed filling degree with the reinforcement $\varphi = 0.6$ of the main material of the flanges. Technological regimes

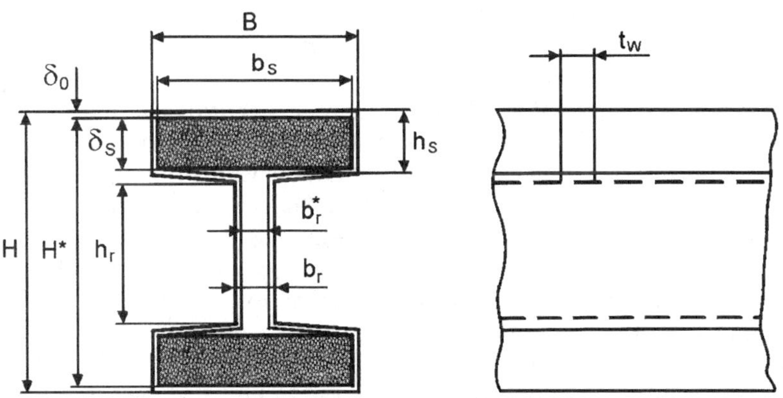

Figure 9.24 Geometrical parameters of the profile.

TABLE 9.5 Geometrical Parameters of the Profiles

Parameters	Design values	Accepted parameters
H^*, mm	28.4	28.8
b_s, mm	14	14.8
δ_s,mm	4.65	4.8
b_r^*,mm	2.4	2.8
δ_0,mm	0.55	0.6
H, mm	30	
B, mm	16	
h_s, mm	6.7	
h_r, mm	15.2	
b_r, mm	4	
t_w, mm	4.4	4

were chosen in accordance with the experimental results presented in the previous chapters to reach continuity, surface finish and required filling degree. After fabrication, the articles underwent additional solidification in a thermal chamber at 180°C for 2 h and thermal stabilization at 22°C for 16 h.

Types of loading used in the tests were chosen to exhibit advantages of the proposed profile structures and their technology most vividly.

Compression tests in the axial direction have disclosed that the reinforcement material in the samples leaning against their faces loses stability under the action of contact stresses, i.e., faces of the samples collapse. To this end, the sample faces were poured in special metal bearing cups. As a result, the samples failed in their working zones restricted by pouring areas.

Results of the tests described are presented in Table 9.6. One conclusion that can be made is that production of complex-profile products using material structures incorporating a braiding layer and stitched joint raises strength parameters of the reinforcing material by 30–40 percent. In addition, this radically alters the failure behavior of the samples.

For instance, samples of a unidirectional material without a braiding layer display a delamination character of failure in longitudinal direction, i.e., they break as separate groups of fibers lose their stability. The samples containing the braiding without the flange joint stitching with the post failed as a result of delamination in only the named joint. As for the samples produced by the developed procedure, they display a characteristic brittle isotropic failure behavior in the direction of the strongest tangential stresses at an angle close to 40–45°.

During bending tests, the unidirectional composite samples failed because of the interlayer shearing in the post area followed by crumpling and breakage of the flange material. Other types of samples broke in

TABLE 9.6 Mechanical Characteristics of Pultruded Profile Elements of Carbon Plastics

	Unidirectional carbon plastics, $\alpha = 0$		Without braiding, $\alpha_n = 0$, $\alpha_c = \pm 45^\circ$		Braiding without stitching, $\alpha_n = 0$, $\alpha_c = \alpha_0 = \pm 45^\circ$		Braiding with stitching, $\alpha_n = 0$, $\alpha_c = \alpha_0 = \pm 45^\circ$	
Material properties	Mean values	w, %	Mean values	w, %	Mean values	w, %	Mean values	w, %
σ_x^-,MPa	417.0	4.7	500.0	5.8	571.0	6.5	607.0	4.3
σ_u,MPa	636.0	1.0	684.0	8.4	707.0	7.7	730.0	10.1
τ_{xy},MPa	27.6	8.3	41.8	4.2	43.1	3.3	48.5	5.6

the upper flange area under the action of compressive stresses. These results match up the test results obtained at axial compression.

The test results from the short beam method have shown perfect agreement with the results of analogous materials by other test procedures.[141,271,361,363] The profiles of unidirectional materials displayed fracture-induced delamination along fibers, and the others failed as a result of cross-shear in the sheet plane.

9.7 Conclusions

1. Investigations of interlaminar strength of materials produced by the braiding–pultrusion method have visualized its interrelation with technological regimes, based on which efficient values of drawing rates and thermal treatment of the profiles were estimated.
2. As a result of studies of the stress–strain state and bearing capacity of rod articles at long compression, their interrelation with parameters of the braiding layer has been established and recommendations have been worked out regarding the choice of efficient reinforcement angle values and the braiding layer thickness.
3. Based on experimental studies of compressible rods, it was found that the pulling velocity affects essentially the bearing capacity of the rods, and the variability of these velocities has been determined taking into account the geometrical parameters of the products.
4. The proposed method of estimating the bearing capacity of tubular articles under the action of inner pressure helps to qualitatively forecast behavior of articles under this type of loading.
5. Results of comparative experimental analyses of complex-profile wares have confirmed that the developed method and material structure improve mechanical characteristics of the fibrous reinforcing filler by 30–40 percent.

Chapter

10

Application of Rods Produced by Braiding–Pultrusion in Roof Bolting

10.1 Composites for Roof Bolting

Roof bolting was adopted as a practice of the mining industry at the beginning of the 20th century. In contrast to previously used types of supports, roof bolting is still widely applicable today.[210,211,238,270,275,279,315,322,407,408,424] It is more efficient and has the following advantages:

- Raises mining safety by resisting blasting and can be mounted for temporary support in mine faces
- Can be completely mechanized
- Requires less mounting hardware and less expenses on their delivery
- Helps to reduce the excavation cross-section by 18–25 percent and its aerodynamic resistance (in contrast to framed lining)

When the conditions and interaction between the massif and supports,[179,236,237,283,311,424] are considered, the following three main schemes of anchorage can be singled out (Fig. 10.1).

The first scheme (joining) is more applicable and operates as follows. The anchored stratified rock is joined into a single beam until crack generation (Fig. 10.1(*a*)). Its resultant bending strength is higher than the total resistance of separate unfixed layers. As soon as cracks appear, the anchors assist in forming high enough rock blocks that wedge and do not fall into the excavation. For the second scheme (fastening), it is characteristic that failed rocks in the excavation roof are fastened by anchors to undamaged rocks beyond the fractured roof limit, i.e., to

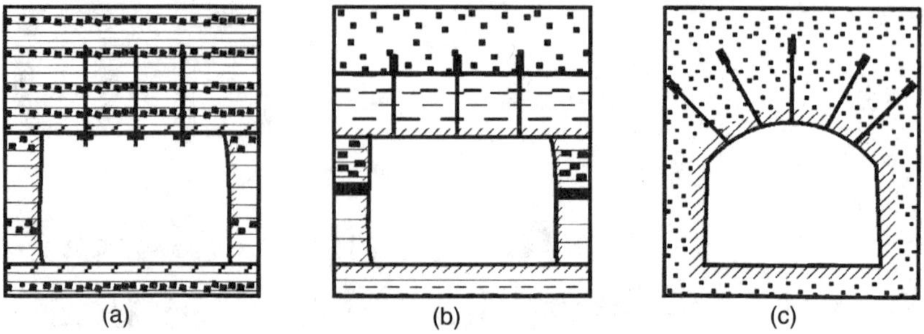

Figure 10.1 Anchorage scheme: (*a*) by joining; (*b*) by fastening; (*c*) for averting rock fall.

strong rocks of the main roof (Fig. 10.1(*b*)). These anchors should strong enough to bear the weight of all rocks in the failed roof or the whole pack of the loaded roof. The third anchoring scheme is used for strong cracked rocks, where the anchorage operates like a walling structure to protect the excavation from accidental falls (Fig. 10.1(*c*)).

Anchorage can be used as a permanent or temporary structure, either individually or in combination with traditional metal or concrete props. Sometimes barb bolts serve for auxiliary purposes to suspend conveyors, monorail systems, pipelines, or cables (Fig. 10.1(*c*)).

Today, modern mining industry employs chiefly rods (anchors) made of rebar or other types of rolled rod-like metal[236,424] instead of reinforcing bolts. These anchors are subjected to a specific machining. Their head is sharpened at a 45° angle or cut along the anchor to a 40–50 mm length, and the tail is threaded over a 50–80 mm area for joining through an adapter with a drill. The rotating anchor is inserted into a borehole, thus creating conditions for required tension after clamping in the borehole.

Service conditions in the rockhole and interactions of the barb bolt with the massif depend much on the method of clamping.[12,138,186] Either the bolt can be clamped rigidly in one point of the rockhole (concrete or polymer lock) or the lock is mounted so that it slips in the borehole (metal stay anchor). The barb bolt design and clamping depend on mining conditions and the anchorage system. Although roof bolting lessens shifts between different points in the zone of inelastic strains, it cannot remove them completely. An anchor tightly clamped over the whole length cannot transfer the tension force onto all rock layers, so there is always a possibility of rupture and loss of bearing capacity. It is therefore advisable to fasten anchors in certain points, thus making them perceive rock shifts lengthwise thanks to slippage of the lock in the rockhole or due to a special structural pliability. Roof bolting will operate most efficiently if service characteristics of the anchor meet geotechnic parameters of rock deformation.

Along with longitudinal tensile loads arising from tightening of the roof bolting and additional operation loads exerted by shifts of the roof layers, the anchor body experiences shear stresses. These shear stresses should be accurately defined because the composites used as bearing elements in such structures display low resistance to shear loads. It is also important to take into account the probability of rod deformation during prolonged loading because of viscoelastic properties of composites. Besides, creep parameters of the structure under preloading should be determined when selecting materials for the anchors.

The major characteristics of the roof bolting are anchor length, density of their arrangement, and tension value. It has been experimentally established[211,231,312,323] that anchor length should not exceed 2–2.5 m for various mining and geological conditions and that the tension is to range within 30–50 kN. The density of installation is determined from the condition of 0.5–1.2 m^2 of the excavation surface area per anchor.

More than 55–65 million different kinds of anchors are used per year in the world.[235,239,424] Most spread types of wedge and end metal anchors are illustrated in Fig. 10.2.

In some cases, wooden anchors find application as short-term, protecting supports.[239,424] They are usually 40–70 mm in diameter round poles with longitudinal face slots from both sides intended for wedges.

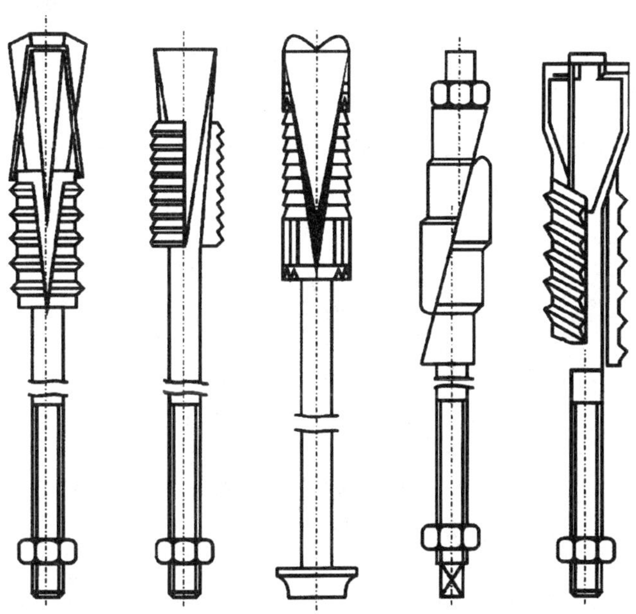

Figure 10.2 Anchor designs.

The slots are mutually perpendicular. It is most expedient to manufacture wood anchors from a hardened compacted or plasticized wood. Application of such anchors is restricted because their bearing capacity is only about 8–12 kN.

Reinforced concrete anchors seem to prevail today in the mining industry and especially in construction of underground structures.[211] They are composed of a cement–sand slurry and steel anchors, and they are highly resistant to shear and tensile loads, and cheap and simple to manufacture. The anchors used in them most often employ steel reinforcement with a periodic profile and 16–24 mm diameter. A fine-grain sand–concrete mix prepared with Portland cement is used as a filler.[332] The bearing capacity of reinforced concrete anchors reaches 80–100 kN, depending on their length. In spite of undoubted advantages, these anchors require special mounting devices and cannot be dismantled.

Lately, reinforced polymer roof bolting[124,177,236,282,382,407,408] fastened into the massif by fixing agents has come into use. Such anchors are inserted into isolated ampoules, which have the properties of rapidly hardening at mixing and strongly adhering to borehole walls and reinforcing rods. Depending on their purpose, technological regimes, and storage conditions, transport and use of several types of ampoule designs have been developed (Fig. 10.3) distinguished by their composition, appearance, and dimensions.

Proceeding from the analysis of roof bolting structures and their potentialities, the following recommendations can be given for ampoule designing:[125]

- Bearing capacity should be on par with the metal roof bolting
- They should resist without any damages the machine torque at installing the roof bolting

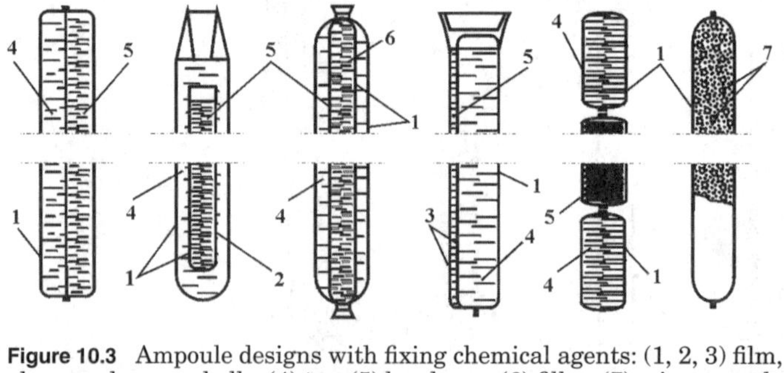

Figure 10.3 Ampoule designs with fixing chemical agents: (1, 2, 3) film, glass, and paper shells; (4) tar; (5) hardener; (6) filler; (7) microcapsule with water.

- The structure is to be easily handled
- They must meet technical and economical norms to ensure efficiency of mining

Some geotechnic conditions require a roof bolting whose material will not resist cutters of excavation and heading machines and can be easily destroyed by blast works and hand tools.

Most fully comply with required demands of the roof bolting,[382] whose anchors are made of polymers that exceed the properties of metals and even surpass them in mechanical properties. The bearing capacity of the roof bolting depends not only on the anchor's durability but also how strongly it is fixed in the borehole as well. A combination of the polymer anchor with fixing chemical compositions yields the best strength characteristics of the joint.

Accident-free operation of heading machines is achieved by the cutting force of a single tooth to a critical load not evoking irreversible deformation or fracture of the cutter. So far, the bearing core of a roof bolting should have a breaking shear point of about 80 MPa. The roof will be reliably supported if the rupture strength of the polymer anchor will be equal or above that of the metal anchor. The rupture strength of the polymer material must surpass 580 MPa in case the metal anchor is 20 mm in diameter and operates under a 180 kN load.

Analysis of the technical literature[172,322,331,332,382] has shown that the glass-reinforced plastics with rupture strength surpassing 1000 MPa and shear strength 60–80 MPa is most fitting for such roof-bolting, although its characteristics at longitudinal shear are rather low. As binders for these rods, most often epoxy, polyester, or other resins can be used. The highest adhesion to glass is shown by epoxy resins when their compositions reach 30–37 MPa. Polyester resins display much lower adhesion to glass, which is only 7.9 MPa. For heavily loaded rods operating under tension, it is expedient to use aluminoboron silicate glass-reinforced fibers consolidated in a rod by a polyester or epoxy binder. The glass fiber content should be about 65–75 percent.

Based on above recommendations, a number of roof bolting designs of polymer materials have been developed (Fig. 10.4). The anchor illustrated in Fig. 10.4a consists of a glass–plastic rod reinforced by a steel wire spiral from one end and terminated into a steel bushing with a nut thread from the other end. The described anchor is 4.5 times lighter than the metal one of the same length. However, it is not expedient to produce such anchors in batches because of large material expenditures and high cost. Another type of anchor (Fig. 10.4b) has been worked out for depth fastening of rock massif (up to 40 m long). It is not prepared beforehand but is erected directly in the borehole or a well.

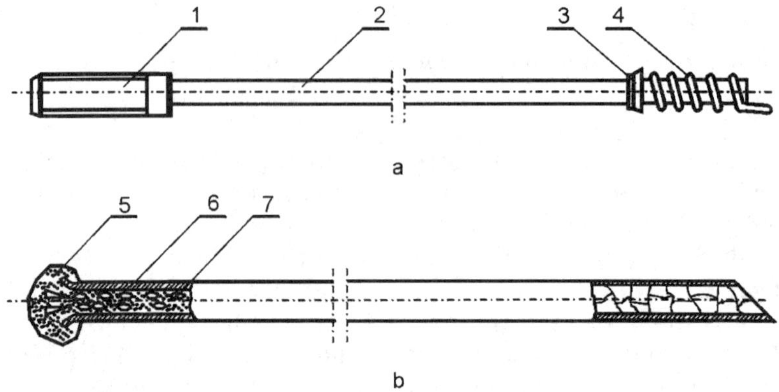

Figure 10.4 Designs of polymer anchor rods: (1) steel bushing; (2) glass–plastic rod; (3) sealing cuff; (4) steel braiding; (5) anchor tip; (6) glass–plastic shell; (7) foam plastic filler.

The anchor represents a glass–plastic tube filled with a foam plastic. Its tip (head) is made of a hard organic concrete and is intended for pressing fastening elements of mines. The inner space of the glass–plastic tube is filled with a few cores of a rigid foam plastic (to the length of the borehole). Such glass–plastic anchors cannot be used individually but only for supporting the existing bolting.

Note that, in spite of certain advantages of proposed anchor rods, the unidirectional composites have a number of essential drawbacks, including weak adhesive bonds in the fastening joint of the rod with the rock, excessive material consumption, low characteristics in transversal direction and at shear, and so on.

In view of the above, some other types of anchor rods have been developed, including a spiral glass–plastic anchor[383] that has elevated cohesive strength with chemical compositions in the borehole due to its developed surface. Such anchors are produced by twisting two strands of glass fiber impregnated with a binder. Their distinctive feature is a load knot made as a loop bending round a steel semi-ring. The anchor material acquires uniaxial orientation and perceives the bending load as in a curved beam. In all other traditional methods of coupling (such as adhesive, thread, rivet, and spline), the tensile load applied is transformed into the shear stress along the layers of the material. The end part of the anchor looks like two braid sections that are coming apart (whiskers), and these break the ampoule in the borehole (see Fig. 10.5).

Application of such twisted anchor rods is restricted because of production complexity, substantial weight, low shear resistance, and low strength and stiffness in the reinforcement direction.

This suggests that the anchor structures described above are far from being adequate for operation in roof bolting. From the analysis of the

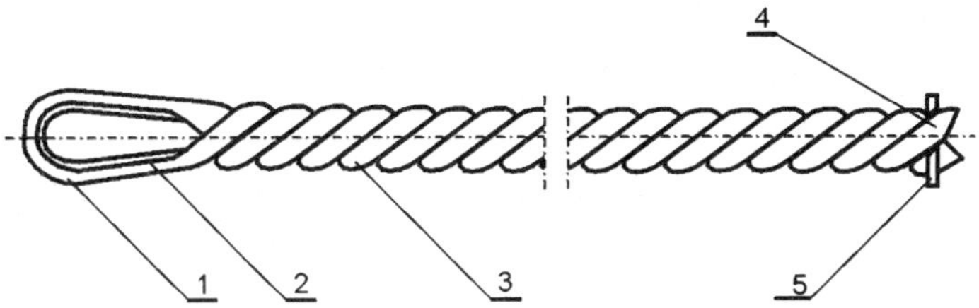

Figure 10.5 Design of a twisted glass-plastic anchor: (1) thimble; (2) steel back; (3) twisted glass–plastic anchor; (4) whisker; (5) split pin.

designs above, it follows that the anchor material should utilize the viscoelastic properties of composites and be ready to resist creep arising from the prestressed state of the anchor at tightening.

Another important requirement of the anchor material and its design is the necessity of organizing mass production at minimal cost and material consumption.

10.2 Materials for Anchor Rods

Attempts to elaborate novel promising designs have led to creation of materials with a spatial reinforcement scheme and usage of multicomponent systems containing different stiffening elements. The new composites successfully compete with traditional materials in cost, surpassing them in mechanical characteristics and quality. New types of fibrous materials with 3D reinforcement networks are composites based on spirally reinforced fillers.[111,112] They are manufactured by spiral winding of one or more kinds of a fibrous reinforcement with organic or glass fibers followed by molding of the polyfibrous braid thus obtained.

Application of additional glass fibers wound at a certain pitch around the braids of continuous parallel fibers, threads, or braids imparts the final product shape stability, impregnation quality, and high filling degree with the reinforcement. As a result, shear and impact resistances improve, and characteristics in the transversal direction are also improved. Use of organic fibers spirally reinforced by glass fibers impregnated with the epoxy binder adds high resistance to the impact loads of the material.[111] Besides, the principle of manufacturing composites with spirally reinforced combined threads and braids widens the assortment of reinforcing fillers, their combinations, and their uses in roof bolting elements.

The problems of strength, stiffness, and stress–strain state of spirally reinforced materials under various types of loading and the effects of structural and geometrical parameters and process regimes on their physicomechanical characteristics have been examined in a number of works.[104,106,107,108,109,110] Furthermore, it has been proved that filling degree of spirally reinforced materials presents a sum of filling with the main φ_{MR} and auxiliary φ_{AR} reinforcements

$$\varphi_0 = \varphi_{\mathrm{MR}} + \varphi_{\mathrm{AR}} \tag{10.1}$$

Note that filling degree with the main reinforcement can be expressed as a product

$$\varphi_{\mathrm{MR}} = \varphi_1 \cdot \varphi_2 \tag{10.2}$$

where φ_1 is the filling degree of the main reinforcement with the fibrous filler and φ_2 is the filling degree of the material with the main reinforcement elements enclosed within a spiral winding.

These two factors are governed by a series of structural and geometrical as well as technological parameters whose efficacy is conditioned by the type of the main reinforcement.

Under the action of the winding braid tension during formation of the filler, there arises a radial constituent of the normal pressure leading to molding of the main reinforcement. The generated pressure value can be calculated by the next formula in presumption of the winding pitch smallness

$$p_{\mathrm{r}} = \frac{2T_{\mathrm{f}}}{D_{\mathrm{r}} t_{\mathrm{w}} \sqrt{1 + \left(\frac{t_{\mathrm{w}}}{\pi D_{\mathrm{r}}}\right)^2}} \tag{10.3}$$

where T_{f} is the tension force of the winding thread, t_{w} is the winding pitch, and D_{r} is the winding reinforcement diameter.

Experimental results of the effect of molding pressure on filling degree of the main reinforcement are illustrated in Fig. 10.6.

The spirally reinforced elements acquire a round shape during winding. So, at rather high filling degree and low pressure values, they tend to preserve this shape. Any changes in this shape at a constant perimeter restricted by the winding may reduce the cross-section and, consequently, increase the filling degree. Examination of produced structures has shown that their shape is close to a circle or ellipse (Fig. 10.7).

An approximate filling degree value of an elliptical element of the main reinforcement can be in a general case found from the relation

$$\varphi_1 = \frac{\varphi_1^0 [1.5(k_1 + 1) - \sqrt{k_1}]^2}{4k_1} \tag{10.4}$$

where φ_1^0 is the filling degree of a round element of the main reinforcement and k_1 is the semiaxis ratio of the elliptical structural element.

The effect of shape variations on the filling degree of the elements of the main reinforcement at different initial filling values is illustrated

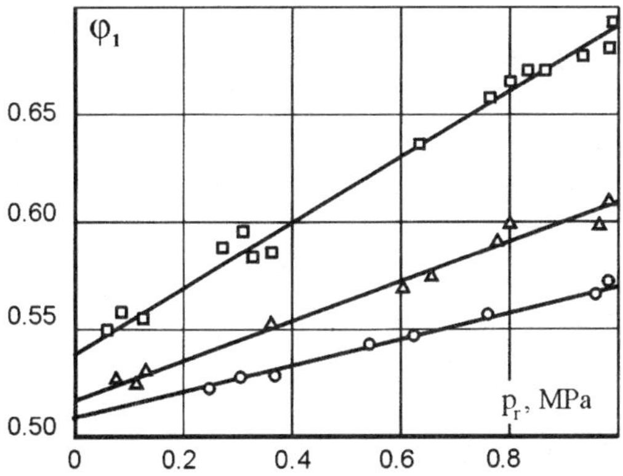

Figure 10.6 Effect of molding pressure on filling degree of the main reinforcement in a spirally reinforced material: (1) at winding a prepreg; (2) wet braids with binder viscosity $\eta = 10$ Pa s; (3) wet braids with binder viscosity $\eta = 4 \cdot 10^{-2}$ Pa s.

in Fig. 10.8. The relative increment of the filling degree with varied k_1 from 1 to 2 approaches 20 percent. In practice, the critical filling degree is restricted by its optimum value, so the inclination of the element to shape variation is restricted as well. Therefore, formula (10.4) can be used to estimate the deformability of structural elements at given filling parameters and optimum filling degrees for a given reinforcement type. Thus we can calculate winding pitch of the reinforcement to ensure a needed deformation.

Let us examine the effect of parameters of a structural element on filling degree of the material with elements of the main reinforcement. In presumption of hexagonal packing of the elements in the material, their elliptical or round shape, and equidistant arrangement given by the winding layer thickness of contacting elements, we can derive a

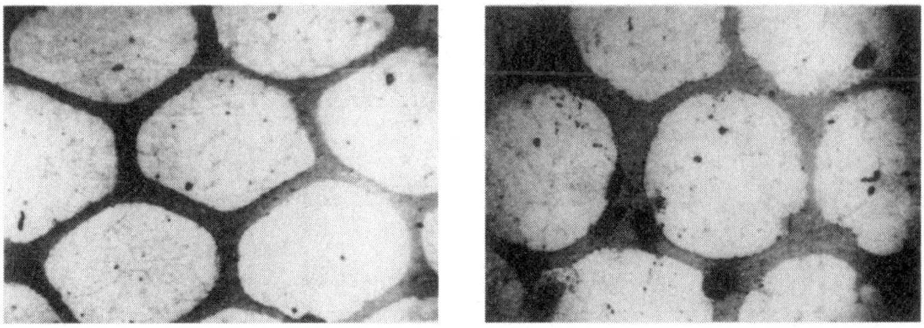

Figure 10.7 Structure of materials based on spirally reinforced fillers.

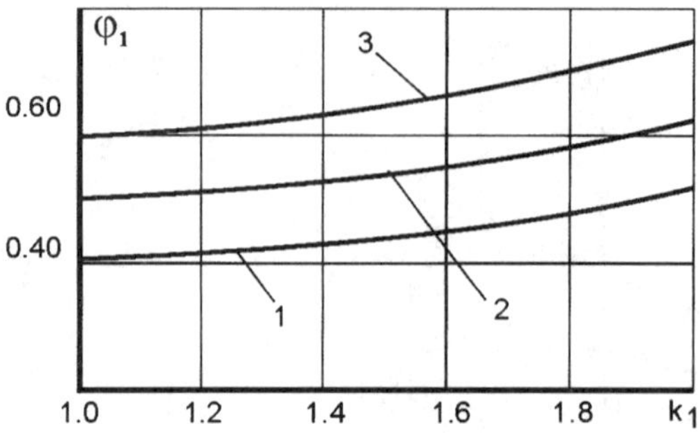

Figure 10.8 Effect of variations in $k_1(\varphi_1^0 = 0.4$ on curve 1, 0.5 on curve 2, and 0.6 on curve 3) on filling degree of the main reinforcement elements at their deformation.

relation of this filling degree value:

$$\varphi_2 = \frac{\pi\sqrt{3}}{6} \cdot \frac{k_1}{(1+k_2)(k_1+k_2)} \tag{10.5}$$

where k_2 is the relation of the distance between elements to their height.

In Fig. 10.9 are shown diagrams of φ_2 dependence on k_1 for different structures. It can be seen that, with incrementing thickness of the winding layer b, the filling degree drops, although under $t_w > 2b/\sin\gamma$ (where γ is the packing layer of the winding reinforcement) this drop is less evident. There is also a region where the tetragonal packing of the elements yields better results than the hexagonal arrangement of the elements with a continuous winding layer.

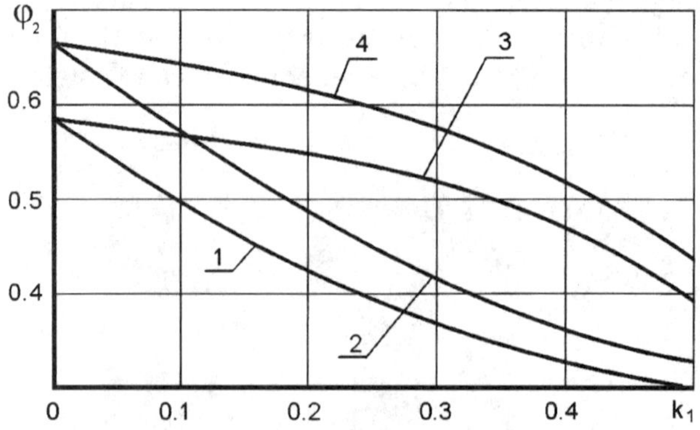

Figure 10.9 Effect of k_1 variations on filling degree of a spirally reinforced material with the main reinforcement: (1, 3) tetragonal packing; (2, 4) hexagonal packing; (1, 2) continuous winding; (3, 4) winding at a pitch $t_w > 2b/\sin\gamma$.

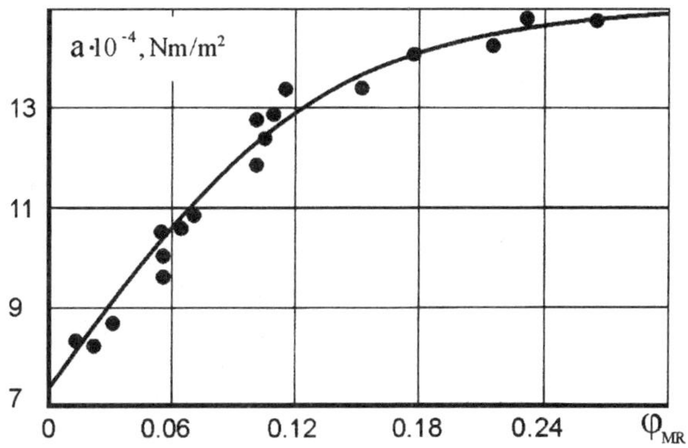

Figure 10.10 Effect of filling degree with auxiliary reinforcement on impact viscosity of a spirally reinforced material.

Hence, when trying to preserve the characteristics of the material in the reinforcement direction, it is most expedient to use in the winding a thin thread at a pitch at which each coil of its winding is laid between the coils of the windings of the neighboring elements. If to meet this condition, the materials with a spiral reinforced filler will gain a 65–75 percent by volume filling with the main reinforcement, which corresponds to keeping elastic characteristics of the material at a former level.

Studies of impact viscosity of considered materials have shown that there is a direct interrelation between specific fracture energy and auxiliary glass reinforcement content in the material (Fig. 10.10). That is why it is, in some cases, necessary to lower the elastic characteristics in the reinforcement direction in order to attain the required impact viscosity parameters.

Investigation results of strength characteristics of the materials based on spirally reinforced fillers are presented in Fig. 10.11 as related to dimensions of a structural element. It follows that an optimum correlation between dimensions of the spirally reinforced elements and the final product diameter can be determined.

where σ_Z^+ = strength of the spirally reinforced material in the main reinforcement direction

σ_{Z0}^+ = strength of the unidirectional material in the reinforcement direction

σ_X^+ = strength of the spirally reinforced material in the transversal direction

σ_{X0}^+ = strength of the unidirectional material in the transversal direction

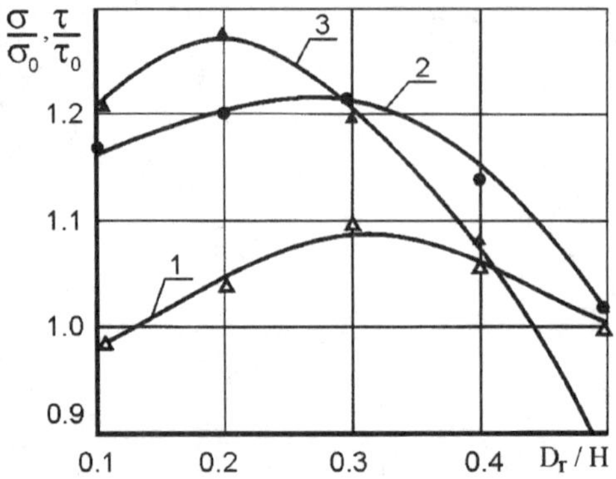

Figure 10.11 Strength dependence on cross-cut ratio to sample thickness: (1) σ_Z^+/σ_{Z0}^+; (2) τ_{XZ}/τ_{XZ0}; (3) σ_X^+/σ_{X0}^+.

τ_{XZ} = strength of the spirally reinforced material at longitudinal shear

τ_{XZ0} = strength of the unidirectional material at longitudinal shear

Analysis of test results shows that increased shear resistance of the materials can be attributed to mechanical meshing of winding coils of the neighboring elements. This effect at cross-separation is connected with enlarged area of adhesion. It should be noted that improvement of strength characteristics in transversal direction is observed only in cases where the cross-sectional size surpasses more than five times the structural element diameter. Otherwise, edge effects exert the effect on the surface and bring about untimely breakage of the material.

Experimental investigations of the strength characteristics of the materials under study have proved that, in fact, under any type of loading they display higher values compared to unidirectional plastics. The winding layer starts to exert its effect on strength characteristics already at the stage of manufacturing spirally reinforced elements. The presence of the winding helps to increase the processing quality of the reinforcement and its contribution into the material. The positive effect is reached as soon as the elementary fibers become straightened, compacted, and the reinforcement braids are reduced. As a result, such components are simultaneously involved in work at loading.

Comparative data on strength characteristics of fillers are cited in Table 10.1. The analysis of presented data shows that winding of the auxiliary reinforcement improves significantly the strength of the filler.

During manufacture of spirally reinforced fillers, the main reinforcement twists to some angle under the torque created by the tension force of the winding thread. The absolute value of the angle depends upon fiber arrangement about the braid center. For the outer layer, it is

TABLE 10.1 Comparative Strength Characteristics of Fillers

Filler type	Relative breaking tension
Binder-impregnated braid	1.15
Dry braid wound with a glass thread	1.39
Binder-impregnated and wound with a glass fiber braid	1.49

about 1–5°. It is indicated in Refs. 181 and 333 that just this twist angle value assists in elevating strength characteristics of the composite materials, thanks to the so-called "cable effect" of the reinforcing braids.

Based on the analysis of crack generation and breakage of composites, it was reported in Ref. 351 that an increased diameter of reinforcing fibers may result in increasing the force required for their pulling out from the material structure. Increased diameter of the fibers may, however, impair their processability and raise damageability, so thus far it is better to group them into bunches. As a result, they preserve flexibility and behave as separate fibers to some extent. It is also indicated that resistance to cracking of such composites was much improved and could be varied by changing mechanical properties of the bunch–matrix interfaces. Such fiber bunches ruptured within the crack planes. It was stated in Refs. 117 and 183 that redistribution of stresses in the fibers at transition to such failure types is roughly an order of magnitude more efficient than at pulling out.

The cable phenomenon is realized in the composite structure at introduction of the auxiliary reinforcement, a result of which the mechanical characteristics at interfaces between the bunches become controllable. Failure types of different samples at tension tests in the direction of the main reinforcement packing are illustrated in Fig. 10.12. Figure 10.12(*a*) presents the materials with a spirally reinforced filler having perfect bonds at fiber bunch interfaces of the main reinforcement (fiber diameter, 0.7 mm; winding pitch, 1 mm with mutual meshing of the coils). The material is seen to preserve its bearing capacity to a certain concentration of broken fibers per unit area, and it breaks without any delamination, which corroborates with conclusions presented elsewhere.[117]

Analysis of the fractured surface has visualized that the amount of fibers pulled out is far less and that their length is about 20–25 μm. Failure of the main mass of fibers is observed in the plane of crack propagation. In contrast, the unidirectional composite breaks as a result of accumulation of damages in various weak or defective sites which is followed by crack propagation to fiber interfaces in the longitudinal direction. Their failure is accompanied by pullout of separate fibers, their groups, or bunches from the structure (Fig. 10.12(*b*)).

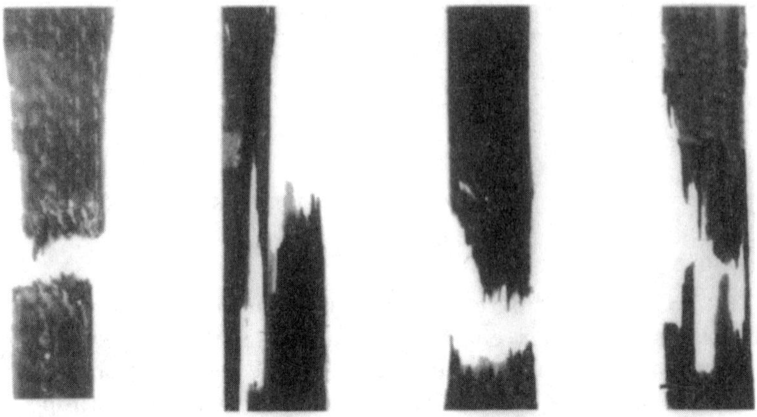

a b c d

Figure 10.12 Failure of samples at tension testing in the main reinforcement direction.

These differences in failure behavior of the samples with continuous and meshing winding of the main reinforcement are illustrated in Figs. 10.12(*a*) and (*c*). In a small, spirally reinforced element (0.7 mm in diameter) with continuous winding, damages are accumulated over a protracted area of the sample because of its low shear resistance. As a result, different elements of the sample break at different portions and are pulled out from the structure. The spirally reinforced samples with larger dimensions (up to 1.6 mm in diameter) acquire larger bunches, and shear character of failure is observed just as inside the element along the contact boundaries of the main and auxiliary reinforcement (Fig. 10.12(*d*)). Such effects are utilized to raise the strength efficiency of the initial reinforcement and to fully or partially eliminate the strength reduction at the expense of diminishing filling degree with the main reinforcement.

As can be seen, use of spirally reinforced fillers leads to redistribution of microstresses in the materials structure and, consequently, to change in failure behavior. Composition of the material and its strength characteristics can be optimized by varying structural, physical, and mechanical properties of its components.

Results of investigations have justified the expediency of using above considered fillers in the structure of tie bolts. This structure is to include a spirally reinforced filler packed in the longitudinal direction along with an outer braided layer that protects the inner structure from damage.

The developed production process of anchor bolts is based on above indicated investigation results and requirements.

10.3 Production Process of Anchor Rods

A combined braiding–pultrusion process has been developed to maintain a continuous production of the rods with previously described complex structure. It consists of a preparatory stage of the spirally reinforced semifinished filler, application of the outer braided layer, and pultrusion of the article in a heated die.

10.3.1 Manufacture of a spirally reinforced filler semi

The additional stage of spiral winding of the main reinforcement involves a number of process parameters. The most important of these parameters are the drawing rate of the main reinforcement, the rotation velocity of the braider, the tension force of the main reinforcement at winding, and the tension force of the winding thread. Of paramount importance is stability of the winding pitch that ensures the condition of meshing of the coils of neighboring elements. The production layout of the semi is illustrated in Fig. 10.13, and a braider is shown in Fig. 10.14.

The pitch stability is affected during spiral winding by the value and persistence of the tension force of the winding thread. When the tension force is small, the friction forces keeping the winding thread in the equilibrium state on the main reinforcement are also low, so the winding may go off the filler, the braking, and guiding devices. This can be eliminated by application of special winders preserving stable tension force of the winding thread. Aside from the binder for the main reinforcement, it is necessary to calculate the amount of the binder for impregnation of the winding threads for the auxiliary reinforcement.

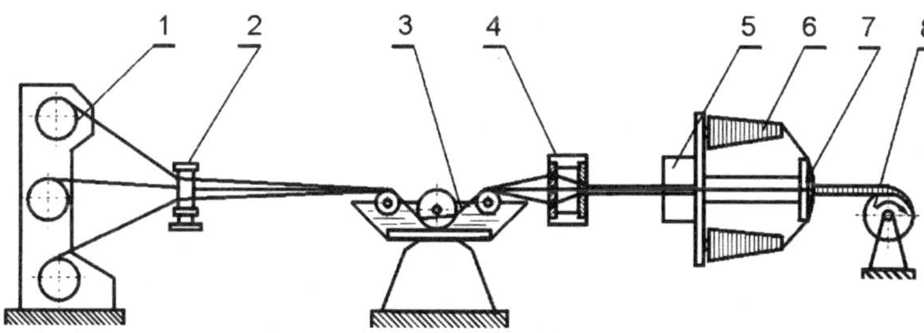

Figure 10.13 Process layout of a spirally reinforced filler: (1) bobbin carrier; (2) strainer of the main reinforcement; (3) impregnation and wet-out unit; (4) combined nozzle; (5) axial coiler; (6) bobbin with auxiliary reinforcement; (7) strainer of auxiliary reinforcement; (8) advance mechanism.

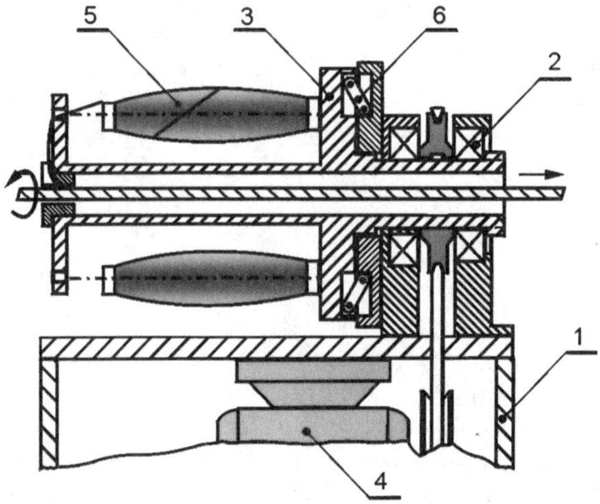

Figure 10.14 Braider design: (1) body; (2) bearing unit; (3) rotor; (4) drive; (5) spool with material; (6) balancer.

The filling mass percentage of impregnated braids is determined by the relation

$$\chi'_{\mathrm{B}} = \frac{\chi_{\mathrm{B}}}{\chi_{\mathrm{B}} + \chi'_{\mathrm{R}} \cdot (1 - \chi'_{\mathrm{L}})} \tag{10.6}$$

where χ'_{R} and χ_{B} are the filling mass percentage of the composite with the main reinforcement and binder; $\chi'_{\mathrm{L}} = m_{\mathrm{L}}/(m_{\mathrm{B}} + m_{\mathrm{L}})$ is the relative solvent content in the ready binder; and m_{B} and m_{L} are the content of the binder and volatile fractions in the solvent.

The amount of the binder necessary for impregnation of the winding layers is calculated by formula (10.6).

Force parameters of the spiral winding are calculated under conditions that preserve linearity of the main reinforcement, as any distortion may result in impaired strength and elastic characteristics.

It has been emphasized in Ref. 114 that the tension force of the winding can be determined based on the condition of preserving linearity of the wound braids

$$P_{\mathrm{w}} \leq \frac{3 \cdot \pi^2 \cdot D_{\mathrm{r}}^3 \cdot [\tau_0]}{16 \cdot t_{\mathrm{w}} \cdot \sin^2 \gamma} \tag{10.7}$$

where $[\tau_0]$ is the limiting stress in the braid resulting from friction and cohesion forces in dry and wet braids and D_{r} is the diameter of the main reinforcement.

Proceeding from the dynamic equation and by giving geometrical and force parameters of the winding process, we can obtain a relation for the critical rotation velocity of the winder

$$n \leq \frac{60}{L}\sqrt{\frac{P_O}{\rho \pi D_r^2}\left(1 - \frac{2P_w C \sin\gamma}{\pi [y] P_O}\right)} \tag{10.8}$$

where P_O = axial tension force of the main reinforcement
$[y]$ = maximum admissible oscillation amplitude
L = distance between poles
ρ = density of the braid material
C = distance from the pole to the point of force application

The proposed procedure of designing the semi with a spirally reinforced filler helps to define the service regimes with a stable spreading of the auxiliary reinforcement.

10.3.2 Molding regimes of outer braided layers

Because the surface layer perceives high shearing loads during pultrusion, it is critical to reach a required degree of interaction between the anchor rod and the rock, for which purpose a layer based on 2D textures is formed. To ensure high, stable properties lengthwise along the material, its porosity should be minimal and filling with reinforcing fibers should be uniform. In addition, the surface density of the braided layer must meet the set requirements,[114,116] and the braids must be reliably fixed on the mandrel surface.

The maximal density of each layer is interrelated with kinematic parameters of the process, such as angular velocity of spindle rotation over the braider guides about the braiding axis and axial feed of the article.

Let us take a coating of a 2D cylindrical braided layer. The involute of the surface is presented in Fig. 10.15.

From triangle ABC we have,

$$\cos\varphi_a = \frac{N \cdot t_b}{\pi \cdot D} \tag{10.9}$$

where N is the number of braider spindles; t_b is the strip width of the braiding reinforcement; and D is the diameter of the cylindrical surface.

Under these conditions, the braided layer is produced with a uniform surface density. Because spindles rotate about the braiding center at a certain frequency n_O (rpm) and drawing velocity V_S (m/min), we can write the reinforcement angle as

$$\tan\varphi_a = \frac{\pi \cdot D \cdot n_O}{V_S} \tag{10.10}$$

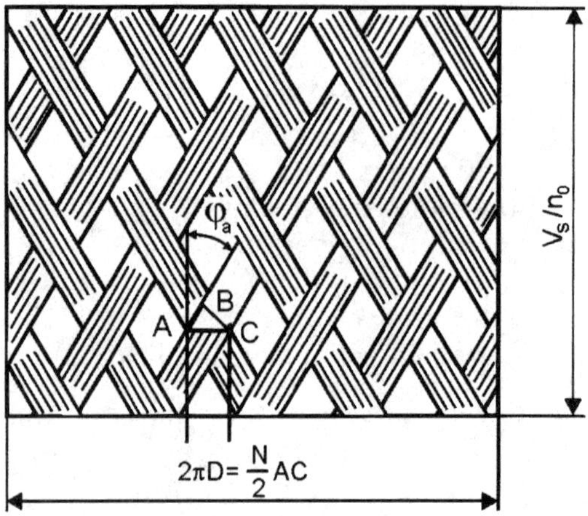

Figure 10.15 Variables considered in the determination of filling degree of a 2D layer.

Taking Eqs. (10.9) and (10.10) into account, the following relation can be derived:

$$\frac{1}{\pi D}\sqrt{\frac{\pi^2 D^2}{N^2 t_b^2} - 1} = \frac{n_0}{V_S} \tag{10.11}$$

If this dependence is obeyed, then the layer on the cylindrical surface will be braided at a constant and maximal given surface density. The design data of the critical braid width dependence on the reinforcement angle and article diameter are shown in Fig. 10.16. Judging by the curves, the material structure becomes overconsolidated when the braid

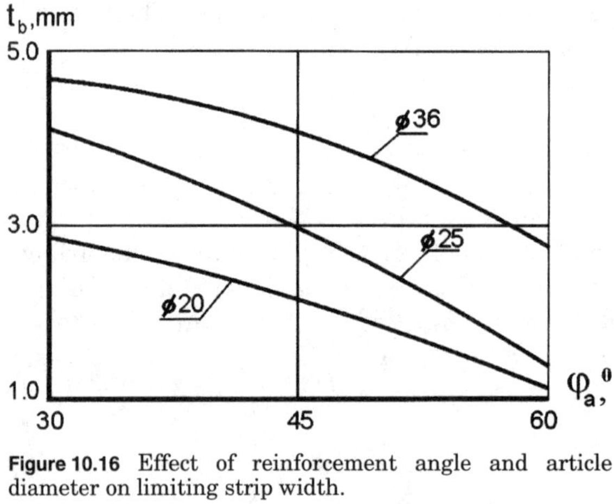

Figure 10.16 Effect of reinforcement angle and article diameter on limiting strip width.

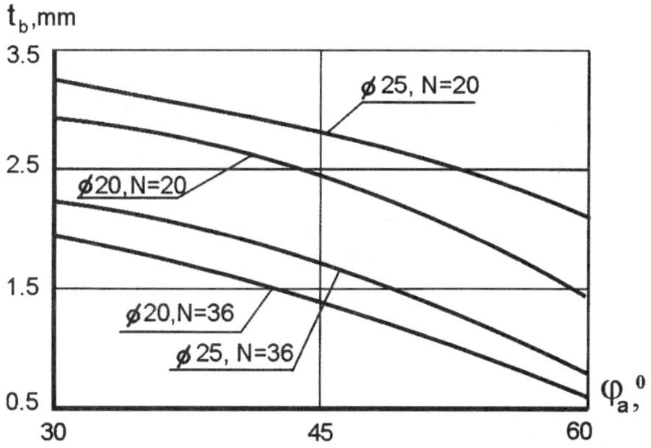

Figure 10.17 Effect of number of carriers and reinforcement angle on the strip critical width.

width is increased above its rated value; otherwise, cells filled only by binder are formed.

Structural parameters can be altered by changing the amount of reinforcement carriers (Fig. 10.17), so the structure will become much more grainy.

So, to achieve a braided layer of a preset surface density and graininess, we can use process (change of the amount of the braiding reinforcement) as well as design (change of the number of carriers) solutions.

10.4 Design of the Anchor Rod Structure

The rod structure must have certain fastening elements on both ends and a device to crush the ampoule with a glue composition. With a tubular element, it is possible to install an inset with a pin whose dimensions are correlated with pressing conditions of the insert into the tubular rod (Fig. 10.18). Pin size is chosen in accordance to the borehole dimensions.

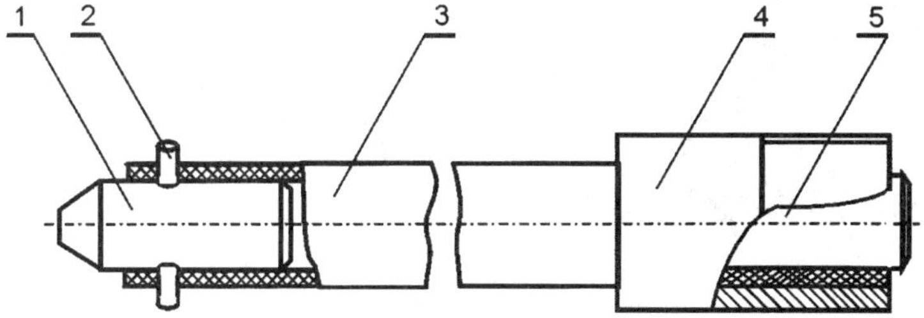

Figure 10.18 Anchor structure: (1) insert; (2) pin; (3) glass-reinforced plastic rod; (4) thread bushing.

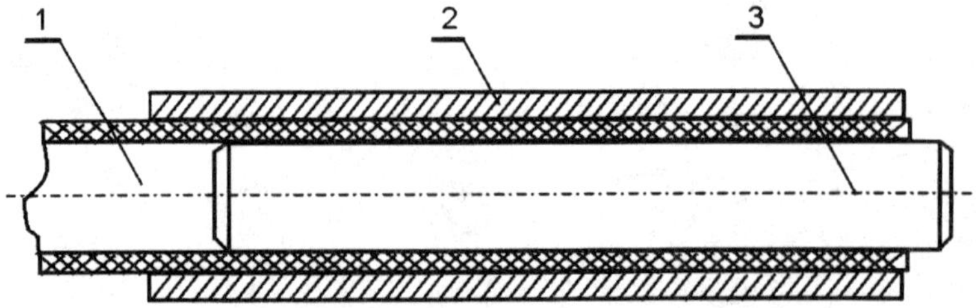

Figure 10.19 Scheme of spacer installation inside the glass–plastic rod: (1) glass–plastic rod; (2) external thread bushing; (3) spacer.

It is important to correctly choose the lower part shape of the anchor rod that transfers the load to the thread bushing. Use of the glued joint over the bushing outer diameter guarantees a certain strength of the junction. It can be calculated based on procedures presented elsewhere.[156,326] However, a stable thickness of the adhesive layer was presumed, and this is difficult to achieve at mass production. In addition, as calculations have shown, the length of the conjugated elements should be rather long, which results in over-expenditure of materials.

To avoid these shortcomings, a junction design with spacers has been elaborated (Fig. 10.19). The inner spacer of a definite diameter was anticipated to exert a considerable pressing force on the glass–plastic tubular rod against the thread bushing. The arising friction force may eliminate the gluing operation and cut the length of the bushing while preserving the reliability of the joint.

To select the dimensions of the glass plastic rod, a problem is solved on its loading with outer and inner pressure (Fig. 10.20) under obeyed boundary conditions on the outer and inner contours of the tube.[16,174,326]

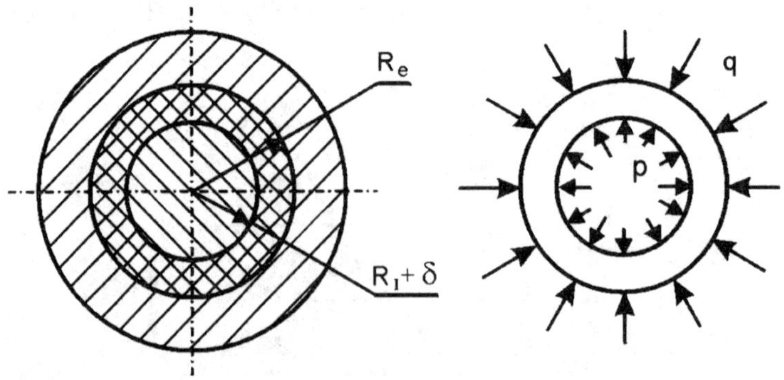

Figure 10.20 Anchor cross-section in spacer locus and loading scheme of the glass–plastic rod.

To calculate stress and strain distributions in the rod wall, let us define the anisotropic parameters[325]

$$K = \sqrt{\frac{E_\Theta(1 - \nu_{rz}\nu_{zr})}{E_r(1 - \nu_{z\Theta}\nu_{\Theta z})}} \tag{10.12}$$

where E_r, $\nu_{z\Theta}$, $\nu_{\Theta z}$, E_Θ, ν_{rz}, and ν_{zr} are the modules of normal elasticity and Poisson ratios of the rod wall in the section plane.

When the tubular rod operates at loading in conditions of plane deformation, the equilibrium equation will be of the type

$$\frac{\partial \sigma_r}{\partial r} + \frac{\sigma_r - \sigma_\Theta}{r} = 0 \tag{10.13}$$

where σ_r and σ_Θ are the radial and tangential stresses and r is the current section radius.

The constraint equation of stress to strain is

$$\begin{aligned} \varepsilon_r &= \frac{\sigma_r}{E_r} - \frac{\nu_{\Theta r}}{E_\Theta}\sigma_\Theta \\ \varepsilon_\Theta &= -\frac{\nu_{r\Theta}}{E_r}\sigma_r + \frac{\sigma_\Theta}{E_\Theta} \end{aligned} \tag{10.14}$$

The constraint equation relating strain to displacements will be

$$\varepsilon_r = \frac{\partial U_r}{\partial r} \qquad \varepsilon_\Theta = \frac{U_r}{r} \tag{10.15}$$

From Eq. (10.15) we obtain

$$r\frac{\partial \varepsilon_\Theta}{\partial r} + \varepsilon_\Theta - \varepsilon_r = 0 \tag{10.16}$$

If we transform Eq. (10.13) into

$$r\frac{\partial \sigma_r}{\partial r} + \sigma_r - \sigma_\Theta = 0 \quad \Rightarrow \quad \frac{\partial (r \cdot \sigma_r)}{\partial r} - \sigma_\Theta = 0$$

and introduce the stress function $\psi = r \cdot \sigma_r$, we arrive, after substitutions, at $\partial\psi/\partial r = \sigma_\Theta$.

To calculate stresses we are to find the stress function φ, for which we transform Eq. (10.16) with Eq. (10.14) into

$$\begin{aligned} & r\left(\frac{1}{E_\Theta}\frac{\partial \sigma_\Theta}{\partial r} - \frac{\nu_{r\Theta}}{E_r}\frac{\partial \sigma_r}{\partial r}\right) + \frac{\sigma_\Theta}{E_\Theta} \\ & \quad - \frac{\nu_{r\Theta}}{E_r}\sigma_r - \frac{\sigma_r}{E_r} + \frac{\nu_{\Theta r}}{E_\Theta}\sigma_\Theta = 0 \end{aligned} \tag{10.17}$$

The relation between elasticity moduli and Poisson ratios, taking into account the orthotropic properties of the rod walls, will be

$$E_r \nu_{\Theta r} = E_\Theta \nu_{r\Theta} \tag{10.18}$$

otherwise, in case of roundness of the spirally reinforced element, we may take

$$\nu_{zr} = \nu_{z\Theta} \qquad \nu_{\Theta z} = \nu_{\Theta r}$$

and from Eq. (10.12) we have $k^2 = E_\Theta / E_r$. Equation (10.14) will then be of the form

$$\nu_{\Theta r} = k^2 \nu_{r\Theta} \tag{10.19}$$

Taking into account Eq. (10.19) and substituting σ_Θ and σ_r expressed through the stress function ψ, instead of Eq. (10.17) we will then have the relation

$$\frac{\partial^2 \psi}{\partial r^2} + \frac{1}{r}\frac{\partial \psi}{\partial r} - \frac{k^2 \psi}{r^2} = 0 \tag{10.20}$$

The solution of this differential equation will be the function ψ, which can be written as

$$\psi = C_1 r^k + C_2 r^{-k} \tag{10.21}$$

where C_1 and C_2 are the constants determined from boundary conditions.

Further, if we substitute Eq. (10.21), we can obtain the following formulas for stresses and radial displacements:

$$\begin{aligned} \sigma_r &= \frac{\psi}{r} = C_1 r^{k-1} + C_2 r^{-(k+1)} \\ \sigma_\Theta &= \frac{\partial \psi}{\partial r} = C_1 r^{k-1} \cdot k + C_2 r^{-(k+1)} \cdot k \\ U_r &= r \cdot \varepsilon_r = \frac{1}{E_\Theta}[C_1 r^k (k - \nu_{\Theta r}) - C_2 r^{-k}(k + \nu_{\Theta r})] \end{aligned} \tag{10.22}$$

When defining constants C_1 and C_2, we anticipate that inner p and outer q pressures are generated on the outer and inner contours of the rod section as the inner spacer is installed. Then, proceeding from the boundary conditions,

$$\begin{cases} r = R_e & \sigma_r = q \\ r = R_I + \delta & \sigma_r = p \end{cases} \tag{10.23}$$

we have

$$\begin{cases} C_1 R_e^{k-1} + C_2 R_e^{-(k+1)} = q \\ C_1 (R_I + \delta)^{k-1} + C_2 (R_I + \delta)^{-(k+1)} = p \end{cases} \tag{10.24}$$

Constants C_1 and C_2 are found from

$$\begin{aligned} C_1 &= \frac{R_e^2 (R_e + \delta)^2}{R_e^2 - (R_e + \delta)^2} (p - q) \\ C_2 &= q - \frac{(R_e + \delta)^2}{R_e^2 - (R_e + \delta)^2} (p - q) \end{aligned} \tag{10.25}$$

Stresses and displacements can be calculated by the relations

$$\sigma_\mathrm{r} = \left(\frac{r}{R_e}\right)^{k-1} \left[\frac{q - p\left(\frac{R_e}{R_\mathrm{I}}\right)^{-(k+1)}}{1 - \left(\frac{R_e}{R_\mathrm{I}}\right)^{2k}}\right] + \left(\frac{r}{R_e}\right)^{-(k+1)} \left[\frac{q - p\left(\frac{R_e}{R_\mathrm{I}}\right)^{k-1}}{1 - \left(\frac{R_e}{R_\mathrm{I}}\right)^{2k}}\right]$$

$$\sigma_\Theta = \left(\frac{r}{R_e}\right)^{k-1} \left[\frac{q - p\left(\frac{R_e}{R_\mathrm{I}}\right)^{-(k+1)}}{1 - \left(\frac{R_e}{R_\mathrm{I}}\right)^{2k}}\right] k - \left(\frac{r}{R_e}\right)^{-(k+1)} \left[\frac{q - p\left(\frac{R_e}{R_\mathrm{I}}\right)^{k-1}}{1 - \left(\frac{R_e}{R_\mathrm{I}}\right)^{2k}}\right] k$$

$$\begin{aligned} U_\mathrm{r} = \frac{1}{E_\Theta} \left(\frac{r}{R_e}\right)^{k-1} \cdot \left[\frac{q - p\left(\frac{R_e}{R_\mathrm{I}}\right)^{-(k+1)}}{1 - \left(\frac{R_e}{R_\mathrm{I}}\right)^{2k}}\right] \cdot (k - \nu_{\Theta\mathrm{r}}) r \\ - \frac{1}{E_\Theta} \left(\frac{r}{R_e}\right)^{-(k+1)} \cdot \left[\frac{q - p\left(\frac{R_e}{R_\mathrm{I}}\right)^{k-1}}{1 - \left(\frac{R_e}{R_\mathrm{I}}\right)^{2k}}\right] \cdot (k + \nu_{\Theta\mathrm{r}}) r \end{aligned} \tag{10.26}$$

Supposing roundness of the spirally reinforced element, we have $k = 1$. We then have the relation for stresses from Eq. (10.26)

$$\begin{aligned} \sigma_\mathrm{r} &= \frac{1}{1 - A^2} \left[q \cdot \left(1 + \left(\frac{R_e}{r}\right)^2\right) - p \cdot \left(\frac{1}{A^2} + \left(\frac{R_e}{r}\right)^2\right)\right] \\ \sigma_\Theta &= \frac{1}{1 - A^2} \left[q \cdot \left(1 - \left(\frac{R_e}{r}\right)^2\right) + p \cdot \left(\left(\frac{R_e}{r}\right)^2 - \frac{1}{A^2}\right)\right] \end{aligned} \tag{10.27}$$

and for displacements

$$U_{\mathrm{r}} = \frac{r}{A^2(1-A^2)E_{\Theta}}\left\{q\left[A^2(1-\nu_{\Theta\mathrm{r}}) - \left(\frac{R_e}{r}\right)^2 A^2(1+\nu_{\Theta\mathrm{r}})\right]\right.$$

$$\left. - p\left[(1-\nu_{\Theta\mathrm{r}}) - (1+\nu_{\Theta\mathrm{r}})A^2\left(\frac{R_e}{r}\right)^2\right]\right\} \tag{10.28}$$

where $A = R_e/R_{\mathrm{I}}$.

To determine pressures p and q, we make use of the boundary conditions

$$\begin{cases} U_{\mathrm{r}} = 0 & \Rightarrow \quad r = R_e \\ U_{\mathrm{r}} = \delta & \Rightarrow \quad r = R_{\mathrm{I}} + \delta \end{cases} \tag{10.29}$$

so, from the system of equations

$$\begin{cases} C_1 R_e(1-\nu) - \dfrac{1}{r}C_2(1+\nu) = 0 \\ \dfrac{1}{E}\left\{C_1(R_{\mathrm{I}}+\delta)\cdot(1-\nu) - \dfrac{1}{R_{\mathrm{I}}+\delta}C_2(1+\nu)\right\} = \delta \end{cases} \tag{10.30}$$

we can obtain pressure values

$$q = \frac{\delta E_{\Theta}}{S_3} \qquad p = \frac{R_e(1-\nu)+S_1}{S_1}q \tag{10.31}$$

where

$$S_1 = 2\cdot\frac{R_e(R_{\mathrm{I}}+\delta)^2}{R_e^2-(R_{\mathrm{I}}+\delta)^2}$$

$$S_2 = \frac{(R_{\mathrm{I}}+\delta)^2(1-\nu)+(1-\nu)R_e^2}{R_e^2-(R_{\mathrm{I}}+\delta)^2}(R_{\mathrm{I}}+\delta)$$

$$S_3 = (R_{\mathrm{I}}+\delta)(1-\nu) - \frac{R_e(1-\nu)+S_1}{S_1}S_2 + S_2$$

Thus, having p and q, we can obtain from Eq. (10.27) the relations for stresses and strains

$$\sigma_{\mathrm{r}} = C_1 + C_2\frac{1}{r^2}$$

$$\sigma_{\Theta} = C_1 - C_2\frac{1}{r^2} \tag{10.32}$$

$$U_{\mathrm{r}} = \frac{1}{E_{\Theta}}\left[C_1(1-\nu)r + C_2\frac{1+\nu}{r}\right]$$

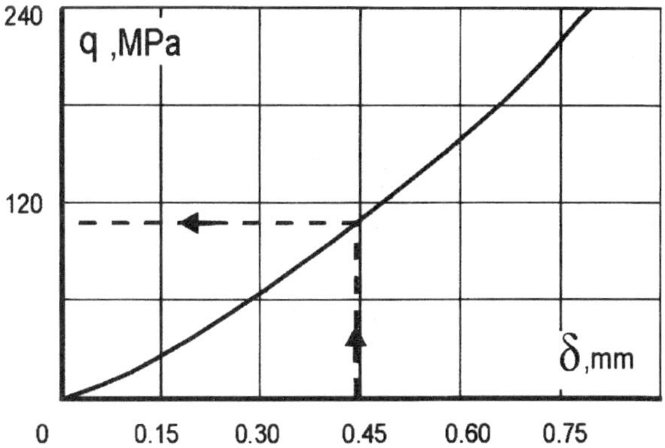

Figure 10.21 Pull effect on outer pressure (arrows indicate critical pressure determination).

The calculation results of the outer pressure q and maximum stress values in the rod wall are illustrated in Figs. 10.21 and 10.22.

To determine maximum admissible spacer size the stresses σ_r and σ_Θ in the rod body were calculated (Fig. 10.23). Changes in strains over the rod radius with different spacer dimensions are shown in Fig. 10.24.

On the basis of the data obtained, we can define the required size of the spacer to ensure the highest acceptable q value under condition of the rod material integrity

$$\sigma_r^{MAX} \leq \sigma_C^- \tag{10.33}$$

where σ_C^- is the breaking compressive strength perpendicular to filler packing.

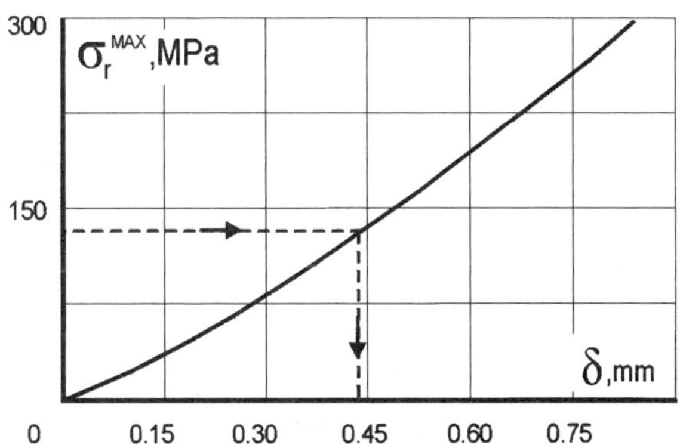

Figure 10.22 Effect of pull increase on maximal radial stresses (arrows indicate the strongest pull determination based on known compressive strength of the rod).

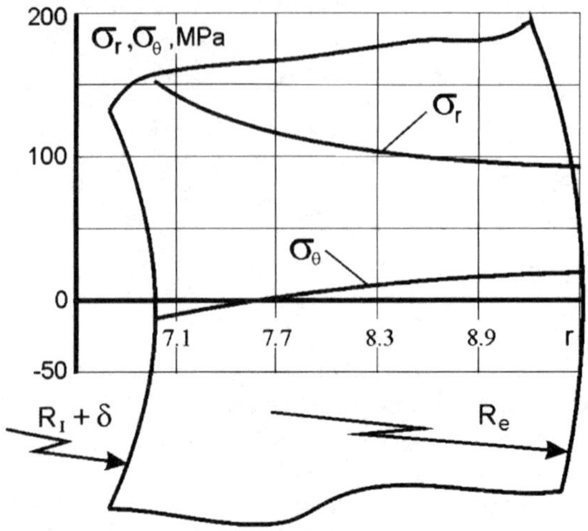

Figure 10.23 Variations of radial and tangential stresses in anchor rod wall ($\varphi = 0.45$).

To find the spacer length, we can use the following relation:

$$L = \frac{P_{\mathrm{MAX}}}{q \cdot \pi \cdot D_e \cdot k_0} \tag{10.34}$$

where P_{MAX} is the maximum load on the anchor rod, D_e is the outer diameter of the anchor rod, and k_0 is the friction coefficient in the glass-plastic–steel pair.

So, according to the proposed method, size of the spacer with account of the loading force required $P_{\mathrm{MAX}} = 18$ kN will be 13.7 cm. Test samples with the thread bushing of length 18 cm and spacer length varying

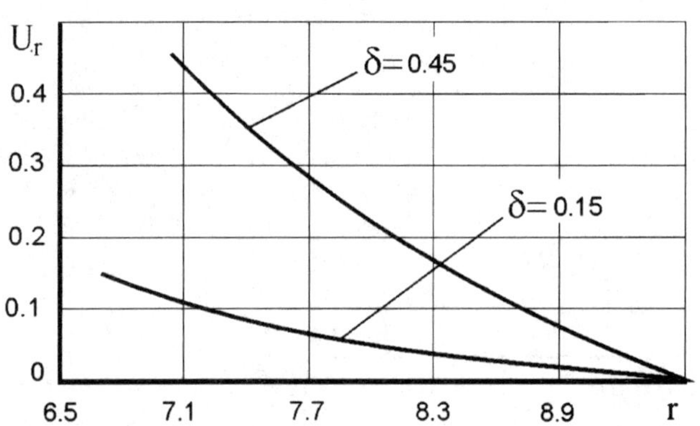

Figure 10.24 Effect of spacer size on deformation of anchor rod wall.

TABLE 10.2 Breaking Force versus Spacer Length

Spacer length, cm	Breaking force, kN
13	14.6–15.0
14	15.3–15.9
15	17.6–18.4
16	18.7–19.1
17	18.7–19.1
18	18.7–19.1

within 13–18 cm were fabricated to verify the theoretical results. Three main types of fracture were observed during the experiments: pull out of a part of the rod from the thread bushing followed by fracture and total failure of the anchor rod outside the fastening zone. The results are presented in Table 10.2.

Analysis of presented data has proved that bearing capacity of the element is reached in case the spacer length somewhat exceeds the rated value. This is evidently because of some eccentricity of the outside and internal surfaces, as a result of which the pressing force of the tubular rode against the inner wall of the thread bushing is distributed unevenly. This is why the spacer length should be extended by 10–15 percent.

As can be seen, the described design of the anchor rod is able to operate reliably under preset loads, and its manufacture is not laborious and economizes costly materials.

10.5 Creep Estimates of the Anchor Rod

Proceeding from the previous case, the bearing anchorage of excavations presupposes a prestrained state of the anchor rod. So far, we must know strain and stress characteristics of the rod and their variations in components of the elements. Creep of a reinforced material depends on the elastoplastic characteristics of the structure and deformation properties of structural components of the anchor.[14,15,52,77,157,174,235,277] Stress values in different components differ and are strongly dependent on the relationship of physicomechanical properties of components and their volume ratio.[10,27,97,102,248,251] As a result of prolonged loading, the stresses in the components start to redistribute with time because of essential differences in elastoplastic characteristics of the polymer binder and reinforcing fillers.[170,230,261]

Elastic constants of the composite and stresses in its components at short-term axial loading in the direction of the main reinforcement packing have been calculated earlier.[113] Stresses perpendicular to the loading direction did not exceed 10–12 percent of those in the main reinforcement direction for the glass-reinforced plastics with spirally reinforced fillers.[107,111] Therefore, these stresses can be neglected for practical aims. The stresses in the main reinforcement direction in the filler and binder were found to be proportional to their stiffness values. For the unidirectional materials, the stiffness ratio in the main reinforcement direction changes at short-term loading within 20–120 depending on fiber type.[11,250] This ratio augments with loading time extension. This suggests that creep-induced deformation of a unidirectional material should be estimated proceeding from deformation properties of its components. It has been demonstrated[9,16,156,247] that this problem is brought to solution by a system of two linear integral equations with account of simultaneous deformations and equations of stress equilibrium. If we substitute these creep kernels of the matrix and reinforcing filler in the form of exponential functions,[17,102] we will have[326]

$$\begin{aligned} \varepsilon_{\mathrm{b}}(t) &= \frac{1}{E_{\mathrm{b}}}\left[\sigma_{\mathrm{b}}(t) + a_{\mathrm{b}}\int_0^t \ell^{-b_{\mathrm{b}}(t-\Theta)}\sigma_{\mathrm{b}}\Theta d\Theta\right] \\ \varepsilon_{\mathrm{f}}(t) &= \frac{1}{E_{\mathrm{f}}}\left[\sigma_{\mathrm{f}}(t) + a_{\mathrm{f}}\int_0^t \ell^{-b_{\mathrm{f}}(t-\Theta)}\sigma_{\mathrm{f}}\Theta d\Theta\right] \end{aligned} \tag{10.35}$$

where $\varepsilon_{\mathrm{b}}(t)$ and $\varepsilon_{\mathrm{f}}(t)$ are the current deformation in the matrix and filler; E_{b} and E_{f} are the moduli of normal elasticity of the matrix and filler; and a_{b}, b_{b}, a_{f}, and b_{f} are the creep parameters of the matrix and filler found from experimental data; correspondingly,[28,167]

$$\frac{a_i}{b_i} = \frac{\varepsilon_i^{\infty}}{\varepsilon_0} - 1 \tag{10.36}$$

where ε_i^{∞} and ε_0 are the corresponding deformations of infinitely long and instantaneous loading.

Thus, creep parameters are calculated as

$$b_i = -\frac{\ln\left(1 - \dfrac{b_i}{a_i}s_1\right)}{[(1+a_i)t_1]^{(1+a_i)}} \tag{10.37}$$

$$a_i = -1 + \frac{1}{\ln\frac{t_1}{t_2}}\ln\frac{\ln\left(1-\frac{b_i}{a_i}s_1\right)}{\ln\left(1-\frac{b_i}{a_i}s_2\right)} \qquad (i = \mathrm{b},\mathrm{f})$$

where $s_1 = [\varepsilon(t_1)/\varepsilon_0] - 1$; $s_2 = [\varepsilon(t_2)/\varepsilon_0] - 1$; and t_1 and t_2 are two independent fixed loading time values.

The additivity rule between stresses is true in any moment in the matrix $\sigma_{\mathrm{b}}(t)$ and filler $\sigma_{\mathrm{f}}(t)$

$$(1-\varphi_{\mathrm{MR}})\sigma_{\mathrm{b}}(t)+\varphi_{\mathrm{MR}}\sigma_{\mathrm{f}}(t)=\sigma_{\mathrm{M}} \tag{10.38}$$

where σ_{M} is constant mean stresses.

Bearing in mind that relative deformation of the whole packet of the material is equal to that of each components,

$$\varepsilon_Z(t)=\varepsilon_{\mathrm{b}}(t)=\varepsilon_{\mathrm{f}}(t) \tag{10.39}$$

we can find from Eqs. (10.35) to (10.38)

$$\begin{aligned}\varepsilon_Z(t)&=\frac{1}{E_Z}\left(1+\frac{a_{\mathrm{b}}a_{\mathrm{f}}+b_{\mathrm{b}}s_3+b_{\mathrm{f}}s_4}{b_{\mathrm{b}}s_5+b_{\mathrm{f}}s_6-b_{\mathrm{b}}b_{\mathrm{f}}}\right)\\&\quad\times\left(1+A_1\ell^{-z_1t}+A_2\ell^{-z_2t}\right)\sigma_{\mathrm{M}}\end{aligned} \tag{10.40}$$

In this case,

$$s_3=a_{\mathrm{f}}\left[1-(1-\varphi)\frac{E_{\mathrm{b}}}{E_Z}\right]\qquad s_4=a_b\left[1-\varphi\frac{E_{\mathrm{f}}}{E_Z}\right]$$

$$s_5=b_{\mathrm{f}}+(1-\varphi)a_{\mathrm{f}}\frac{E_{\mathrm{b}}}{E_Z}\qquad s_6=b_{\mathrm{b}}+\varphi a_{\mathrm{b}}\frac{E_{\mathrm{f}}}{E_Z}$$

$$z_{1,2}=\frac{s_5+s_6}{2}\left(1\mp\sqrt{1-4\frac{s_5b_{\mathrm{b}}+s_6b_{\mathrm{f}}-b_{\mathrm{b}}b_{\mathrm{f}}}{(s_5+s_6)^2}}\right)$$

$$A_1=\frac{s_7-z_1(s_3+s_4)}{(z_1-z_2)(s_7+s_8)}z_2\qquad A_2=\frac{s_7-z_2(s_3+s_4)}{(z_2-z_1)(s_7+s_8)}z_1$$

$$s_7=a_{\mathrm{b}}a_{\mathrm{f}}+b_{\mathrm{b}}s_3+b_{\mathrm{f}}s_4$$

$$s_8=s_5b_{\mathrm{b}}+s_6b_{\mathrm{f}}-b_{\mathrm{b}}b_{\mathrm{f}}$$

Stress variations in material components can be determined with Eq. (10.40) as follows:[156,249,276,326]

$$\begin{aligned}\sigma_{\mathrm{b}}(t)=\frac{E_{\mathrm{b}}}{E_Z}\left(1+\frac{s_7}{s_8}\right)&\left[-a_{\mathrm{b}}\left(\frac{1}{a_{\mathrm{b}}+b_{\mathrm{b}}}+\frac{A_1}{a_{\mathrm{b}}+b_{\mathrm{b}}-z_1}\right.\right.\\&\left.+\frac{A_2}{a_{\mathrm{b}}+b_{\mathrm{b}}-z_2}\right)\ell^{-(a_{\mathrm{b}}+b_{\mathrm{b}})}+\frac{A_1(b_{\mathrm{b}}-z_1)}{a_{\mathrm{M}}+b_{\mathrm{b}}-z_1}\ell^{-Z_1t}\\&\left.+\frac{A_2(b_{\mathrm{b}}-z_2)}{a_{\mathrm{b}}+b_{\mathrm{b}}-z_2}\ell^{-Z_2t}+\frac{b_{\mathrm{b}}}{a_{\mathrm{b}}+b_{\mathrm{b}}}\right]\sigma_{\mathrm{M}}\end{aligned}$$

$$\sigma_{\rm f}(t) = \frac{E_{\rm f}}{E_Z}\left(1+\frac{s_7}{s_8}\right)\left[-a_{\rm f}\left(\frac{1}{a_{\rm f}+b_{\rm f}}+\frac{A_1}{a_{\rm f}+b_{\rm f}-z_1}\right.\right.$$
$$\left.+\frac{A_2}{a_{\rm f}+b_{\rm f}-z_2}\right)\ell^{-(a_{\rm f}+b_{\rm f})}+\frac{A_1(b_{\rm f}-z_1)}{a_{\rm f}+b_{\rm f}-z_1}\ell^{-Z_1 t}$$
$$\left.+\frac{A_2(b_{\rm f}-z_2)}{a_{\rm f}+b_{\rm f}-z_2}\ell^{-Z_2 t}+\frac{b_{\rm f}}{a_{\rm f}+b_{\rm f}}\right]\sigma_{\rm M} \qquad (10.41)$$

On the basis of the relations above, we may estimate the creep values of the unidirectional material in the reinforcement direction, i.e., when characteristics of the binder are substituted in Eq. (10.40), they describe stress variations in the core of the spirally reinforced material.

To evaluate creep in the auxiliary reinforcement, we may consider that of a unidirectional layer under loading in a direction non-coincident with the main reinforcement. In this case, the relative elastoplastic constituent of deformation in the α direction (making an angle α with the main reinforcement direction) will be

$$\varepsilon_\alpha(t) = \overline{S}_\alpha(t)\sigma_{\rm M}^{\rm L} \qquad (10.42)$$

Here $\overline{S}_\alpha$ is the viscoelastic pliability of the unidirectional layer in the axes turned by angle α about the reinforcement direction.

Relation (10.42) can then be presented in the form

$$\varepsilon_\alpha(t) = [a_1 + b_1 D(t)]\sigma_{\rm M}^{\rm L} \qquad (10.43)$$

where

$$a_1 = \frac{1}{E_1}\cos^4\alpha + d_1\sin^4\alpha + \left(2\frac{\nu_{12}}{E_1}+g_1\right)\sin^2\alpha\cos^2\alpha$$
$$b_1 = (d_2\sin^2\alpha + g_2\cos^2\alpha)\sin^2\alpha$$

and E_1 and ν_{12} are the elasticity module and Poisson ratio in the axes coinciding with reinforcement direction.

$$g_1 = \frac{2}{3}(1+\nu_{\rm b})\left(\frac{b_{\rm b}}{G_{12}} - \frac{1}{G_{12}^{(\infty)}}\right) + \frac{1}{G_{12}}$$
$$g_2 = \frac{2}{3}b_{\rm b}(1+\nu_{\rm b})\frac{(G_{12}^{(\infty)} - G_{12})G_{\rm b}}{G_{12}^{(\infty)}G_{12}}$$

where $G_{\rm b}$ = shear modulus of the matrix material
G_{12} = shear modulus of the unidirectional plastic

$[G]_{12}^{(\infty)}$ = factor determined with account of the type of fiber arrangement in the layer and variations in the matrix properties under prolonged loading

d_1 and d_2 = factors determined in Refs. 9, 156, and 326 depending on elastic properties, volume ratio, and arrangement of the components

Hence, deformation of the core and auxiliary reinforcement are described by Eqs. (10.40) and (10.43), respectively. Simultaneously, stresses in layer $\sigma_{\mathrm{M}}^{\mathrm{L}}$ and in the core $\sigma_{\mathrm{M}}^{\mathrm{k}}$ are in equilibrium in any moment

$$(1-m)\sigma_{\mathrm{M}}^{\mathrm{L}} + m\sigma_{\mathrm{M}}^{\mathrm{k}} = \sigma_{\mathrm{M}} \qquad (10.44)$$

where $m = F_{\mathrm{L}}/F_{\mathrm{k}}$ is the cross-sectional area of the auxiliary reinforcement layer ratio to the total area of a representative element of the material structure.

Presuming that the cores of spirally reinforced elements are evenly distributed in the pseudo-matrix whose characteristics are governed by those of the interlayer and based on Eqs. (10.35), (10.36), (10.37), and (10.43), we can draw a curve for the reinforced rod creep at loading along its axis

$$\varepsilon_{\mathrm{r}}(t) = \frac{1}{E_{\mathrm{r}}}\left(1+\frac{s_7}{s_8}\right)\left(1 + A_1^{*}\ell^{-Z_1^{*}t} + A_2^{*}\ell^{-Z_2^{*}t}\right)\sigma_{\mathrm{S}} \qquad (10.45)$$

where $\varepsilon_{\mathrm{r}}(t)$ and E_{r} are the relative strain and elasticity module of the rod; σ_{S} is the operating tensile stress. Factors marked by * can be found from Eqs. (10.36), (10.37), and (10.39).

To accelerate processing of experimental results, we can use a simplified dependence

$$\varepsilon(t) = \varepsilon_0(1+t^n) \qquad (10.46)$$

where $\varepsilon(t)$ = complete deformation

ε_0 = initial strain corresponding to a given load

n = constant determined from

$$n = \frac{\ln\left(\frac{2\varepsilon_2-\varepsilon_1}{\varepsilon_1}\right)}{\ln t_2} \qquad (10.47)$$

ε_1 and ε_2 = strains corresponding to two fixed loading time values t_1 and t_2.

Analysis of the relation above shows that the difference between values obtained by Eqs. (10.45) and (10.46) is below 8 percent.

To estimate the pseudo-matrix properties, experimental investigations have been carried out of creep in tubular rods simulating structural

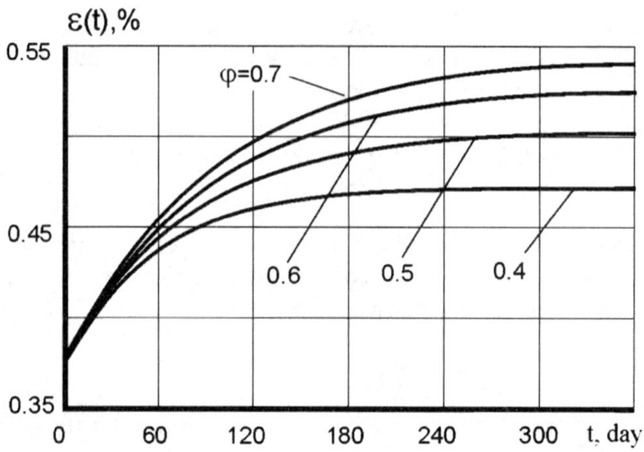

Figure 10.25 Loading time effect on relative longitudinal deformation for various filling degrees of the auxiliary layer and $\alpha = 75°$.

parameters of the layer. The rods were produced by winding glass fibers impregnated with a binder over a cylindrical mandrel. Two main parameters were varied, namely reinforcement angle and the layer filling degree. The recorded parameters were relative longitudinal deformation at initial loading, at 5-day intervals over the course of a year. No fewer than 8 samples were tested to achieve reliable data.

Test results of the samples with different filling degrees at prolonged tension ($\sigma_S = 50$ MPa) and load application at angle $\alpha = 75°$ to the reinforcement direction are presented in Fig. 10.25. Analysis of presented data shows that the approximating curves, based on dependencies (Eq. (10.35)) and taking into account the glass fibers and binder characteristics, adequately describe the deformation process of the material. As is seen, the variation of the filling degree does not perceptibly affect the creep behavior because, under rather large angles between the reinforcement direction and loading, the decisive role in creep behavior is played by shear characteristics of the binder.

To estimate the effect of reinforcement angle on material properties, the samples were made of a prepreg whose width was chosen so as to make the reinforcement angle equal to 60–85°. The filling degree as most efficient and simply attainable during the process value was kept constant $\varphi_0 = 0.6$.

The test results presented in Fig. 10.26 prove that contraction of angle α leads to diminished longitudinal deformation; the difference between the largest strain values under $\alpha = 85°$ and $\alpha = 60°$ is

$$\frac{\varepsilon(\infty)_{\alpha=85°}}{\varepsilon(\infty)_{\alpha=60°}} = 1.09 \tag{10.48}$$

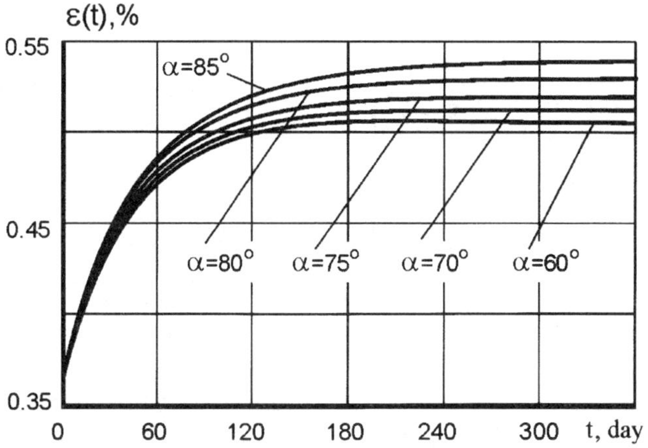

Figure 10.26 Relative deformation of the samples under loading at α angle to reinforcement direction ($\varphi = 0.6$, $\sigma_S = 50$ MPa).

It should be emphasized that, even though this value is insignificant, one should remember that the strains under consideration are referred to as the pseudo-matrix properties, which are critical for characterizing the rod as a whole.

In Fig. 10.27, relative variations of deformation in time are shown for two values of angle $\alpha = 75°$ and $\alpha = 85°$ and filling degree $\varphi = 0.6$. It should be noted that the approximation curves constructed on the basis of above cited data comply well with experimental evidences. The statistical analysis of test results and their comparison with theoretical data has shown that the difference is below 13–16 percent.

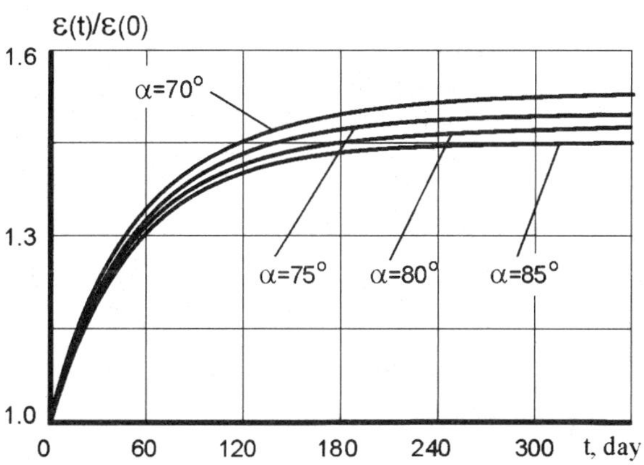

Figure 10.27 Relative variations in longitudinal deformation within loading time at $\varphi_0 = 0.6$ and different α values.

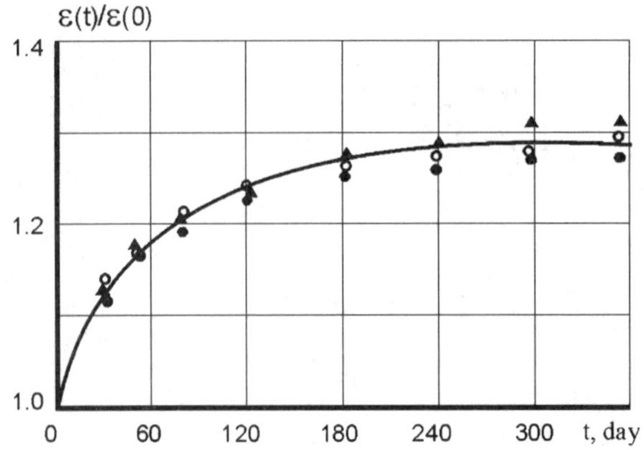

Figure 10.28 Relative variations of longitudinal deformation along reinforcement direction under three stress values ($\varphi_0 = 0.7$).

Deformability of the composite under loading along the reinforcement has been estimated, too. The results of such investigations can characterize the spirally reinforced element cores. To this end, tubular rods with a unidirectional reinforcement scheme (along the axis) were fabricated. Process regimes corresponded fully to that of manufacturing of above described anchors. The tubular samples were cut to length sections and fastened in grips of a testing machine. The tensile load was varied so as to reach 200, 250, and 300 MPa stresses in the sample cross-sections that correspond to anchor tightening at mounting. The test results are illustrated in Fig. 10.28.

The derived experimental data (Figs. 10.27 and 10.28) were used to estimate creep of the material based on the spirally reinforced filler. In Fig. 10.29, both experimental and theoretical results are presented of the anchor rod creep under the three tensile load values. The maximum relative deformation at prolonged tension along the axis of the spirally reinforced filler based rod is seen to be 27 to 35 percent lower than that of the unidirectional material.

Stress variations in the material structure under long-term loading have been estimated using dependencies (Eq. (10.41)), as well as experimental data obtained at loading under various angles to reinforcement directions. The results illustrated in Table 10.3 include relative variations of stresses in the core of the spirally reinforced element σ_k and auxiliary layer σ_L at the initial moment of loading and 400 days after the loading. The stress values have been determined with account of nonuniformity of the auxiliary layer thickness. It is seen that stresses in the auxiliary reinforcement tend to relax with time, whereas those in the main reinforcement layer grow.

TABLE 10.3 Stress Values in Structural Elements of the Rod

Layer thickness	$\frac{\sigma_L(t=0)}{\sigma_S}$	$\frac{\sigma_L(t=400)}{\sigma_S}$	$\frac{\sigma_k(t=0)}{\sigma_S}$	$\frac{\sigma_k(t=400)}{\sigma_S}$
0.05	0.1394	0.0974	1.0882	1.0925
0.10	0.1509	0.1058	1.1783	1.1878
0.15	0.1627	0.1146	1.2700	1.2856
0.20	0.1746	0.1235	1.3630	1.3860

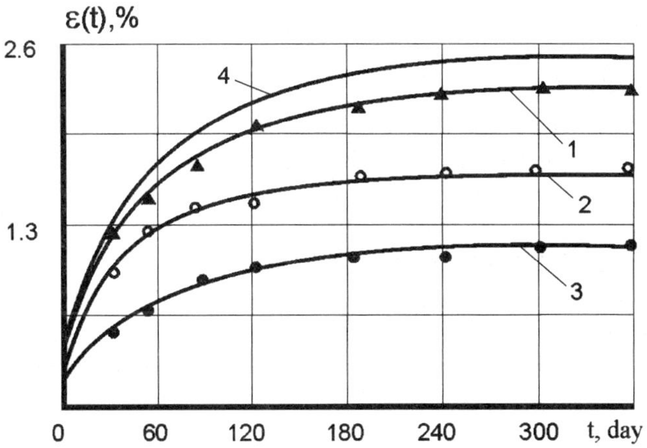

Figure 10.29 Creep of reinforced rods under various tensile stresses: (1, 2, 3) based on spirally reinforced material; (4) unidirectional; (1) $\sigma_s = 300$ MPa; (2) $\sigma_s = 250$ MPa; (3) $\sigma_s = 200$ MPa; (4) $\sigma_s = 300$ MPa.[326]

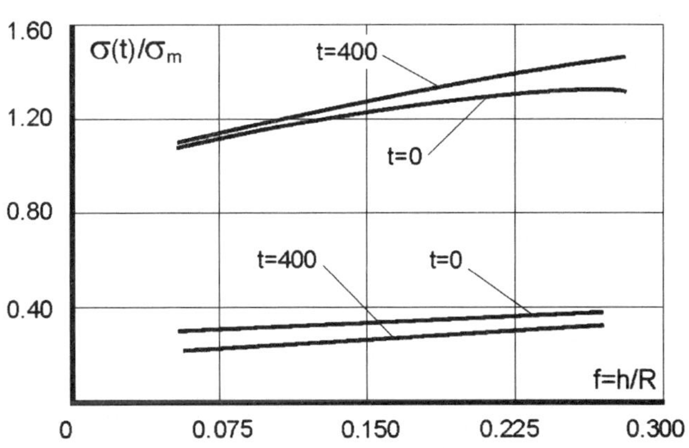

Figure 10.30 Stress variations in components of spirally reinforced rod as dependent on auxiliary reinforcement thickness.

Increased thickness of the auxiliary reinforcement may also lead to redistribution of stresses within the structural elements (Fig. 10.30).

Proceeding from the investigations, a conclusion has been made that application of spirally reinforced fillers lowers the creep of anchors by

25–35 percent and extends service life of the roof bolting without any additional tightening of the anchorage.

10.6 Conclusions

1. To cut material consumption, improve adhesive contact of the rods with the rock and their fixing in the borehole, and constrain shear loads of the rock beds, it was proposed that spirally reinforced fillers be employed in the structure of anchor rods, which creates a 3D reinforcement network and significantly improves the characteristics of the part as a whole.
2. A technological process of manufacturing anchor rods with a complex reinforcement scheme incorporating spirally reinforced fillers has been elaborated based on investigation results. The productivity of the process has been increased by up to 0.6 running meter/min under correctly chosen efficient process regimes.
3. Based on the developed calculation method of fastening units for the anchor rod, a design model able to withstand high operation loads has been elaborated. The material capacity and labor intensiveness of its production have been minimized.

Conclusions and Recommendations

The results of investigations provide the ground for the following conclusion: since rod-like elements of structures experiencing high mechanical loads during operation necessitate rather intricate reinforcement schemes, it seems most expedient to use a combined technological process—one that joins braiding and pultrusion operations as well as stitching in case it is necessary to combine complex-profile articles into a single process. This measure, in addition to an optimum choice of process parameters, will guarantee the required structural, geometrical, and physico-mechanical characteristics of products in the course of the continuous process.

The results of combined investigations of the main process stages are presented, and principles of selecting efficient design and technological solutions of the chief elements of the equipment are proposed.

Based on the investigation results and the existing methods of manufacturing profile composite products, the following conclusions and recommendations have been formulated:

1. A continuous production process of complex-profile articles has been devised as an elaboration on the basic braiding-pultrusion method. It also includes a series of operations of pressure treatment and stitching aimed at the manufacture of products with intricate structure and high-service characteristics.
2. Analytical dependence of process parameters from structural and geometrical characteristics of the material has been established, based on the analysis of kinematics of the braiding process, which assist in manufacturing a material with the maximum possible surface density of the braided layers.
3. Investigations of pultrusion regimes and their effect on physico-mechanical characteristics of produced materials have provided the basis for determining the most efficient parameters of the continuous pultrusion-braiding process, and accordingly, for attaining the required setting degree of the binder and properties of the material.
4. Force parameters of the braiding-pultrusion process and their variations within the range that guarantees the needed structure and

physico-mechanical characteristics of the final product have been estimated.

5. Particular designs for the individual elements of the braiding-pultrusion equipment have been described; these designs will result in the efficient production of a wide range of rods.
6. A method of selecting the most practical applications of the braiding-pultrusion process is proposed, taking into account both the properties of the materials used and their continuous molding.
7. The conducted experimental investigations of physico-mechanical characteristics and microstructure of the materials have established that the proposed production process can be justifiably used for the manufacture of complex-profile, high quality composite ware preset properties.
8. Based on theoretical and experimental studies of the stitching process of high-modular fibers structures, a new kinematic-force design method has been devised. This method makes it possible to quantify the forces and pressures generated by the stitching of braided composite structures, to estimate the optimum stitching regimes of complex-profile articles, and to recommend equipment designs for a given operation.

References

1. Aboudi, J. (1991) *Mechanics of Composite Materials: A Unified Approach*, Elsevier.
2. Abrate, S. (1994) Optimal design of laminated plates and shells, *Composite Structures*, 29, 269–286.
3. Adams, D.F., Odom, E.M. (1987) Tensile and compression test methods for cylindrical weave carbon—carbon composite materials, *SAMPE Quart*, 18, 2, 25–31.
4. *Advanced Composites Manufacturing*. (1997) Timothy Gutowsky (Ed.), John Wiley & Sons.
5. Ahmadian, R., Mantena, P.R. (1996) Modal Characterization of Portal Frames Made of Mechanically Jointed Pultruded Flat Hybrid Composite Beams, *Composites Engineering*, part B, 27B, 319–328.
6. Alfutov, N.A., Zinoviev, P.A., Popov, B.G. (1984) Design of multilayered plates and shells of composite materias, Moscow, *Mashinostroenie*, 64.
7. Allah, M.H.A., Selmy, A.I., Khashaba, U.A., Abdin, E.M. (1996) Effect of fiber volume fraction on the fatigue behavior of GRP pultruded rod composites, *Composites Science and Technology*, 56, 1, 23–29.
8. Alper, J.M., Wang, A.S.D. (1981) Compressive strength of a unidirectional graphite-epoxy composite following a tensile proof-test, *Composite Technology Review*, 3, 156–157.
9. Altenbach, H.A. (2001) Nonclassical model for creep-damage processes, *Mater. Phys. Mech.*, 3, 25–35.
10. Altenbach, H. Creep-damage behavior of plate and shells (1999) *Mechanics of Time-Dependent Materials*, Dordrecht, 3, 103–123.
11. Altenbach, J., Altenbach, H., Naumenko, K. (1997) Lebensdauerabschätzung dünnwandiger Flächentragwerke auf der Grundlage phänomenologischer Materialmodelle für Kriechen und Schädigung, *Technische Mechanik*, Magdeburg, 17, 353–364.
12. Althouse, R.C., Grayson, R.L., Klishis M.J. (1997) Microanalysis of roof bolter injuries, *Journal of Applied Occupational and Environmental Hygiene*, 12, 12, 851–857.
13. Anifantis, N.K. (2000) Micromechanical stress analysis of closely packed fibrous composites, *Composites Science and Technology*, 60, 8, 1241–1248.
14. Aniskevich, K., Yanson, Y.O. (1994) Change in the viscoelastic properties of an organic-fiber-reinforced plastic during loading, *Mechanics of Composite Materials*, 30, 1, 84–89.
15. Aniskevich, N., Yanson, Y.O., Aniskevich, N.I. (1992) Creep of epoxy binder in a humid atmosphere. *Mechanics of Composite Materials (N.Y.)*, 28, 1, 12–18.
16. Arnold, W.S., Marshall, I.H., Wood, J. (1990) Optimum design consideration for mechanically fastened composite joints, *Compos. Struct.*, 3, 85–101.
17. Askadsky, A.A. New probable types of relaxation kernels (1987) *Mech. Comp. Materials*, 3, 403–409.
18. ASTM C364 (1995) Standard test method for edge with compressive strength of sandwich constructions, American Society for Testing and Materials, *Annual Book of ASTM Standards*, 03. 01.
19. ASTM Standard D 2256 (1995) Standard Test for Tensile Properties of Yarns, American Society for Testing and Materials, *Annual Book of ASTM Standards*, vol. 7.01.

20. Avva, V.S., Sadler, R.L., Shivakumar, K.N., Dunn, D., Jessup, S., Ingram, C. (1997) Comparative studies on laminated and textile composites, *AIAA Paper*, 97–1322.
21. Avva, V.S., Shivakumar, K.N., Sadler, R.L., Morgan, B. (1996) Mechanical properties of 3D braided composite panels fabricated using RTM techniques, *AIAA Paper*, 96–1534.
22. Avva Vishnu, S., Alien, H.G. (1996) Through-the-thickness tension strength of 3D braided composites, *Compos. Mater.*, 30, l, 51–68.
23. Aylward, L., Douqlas, C., Roylance, D.A (1985) Transient finite element model for pultrusion processing, *Polymer Process Engineering*, 3, 3, 247–261.
24. Babaevsky, P.G. (1980) Practical works on polymer materials science. Moscow, *Khimiya*, 235.
25. Bakis, C.E., Bank, L.C., Brown, V.L., Cosenza, E., Davalos, J.F., Lesko, J.J., Machida, A., Rizkalla, S.H., Triantafillou, T.C. (2002) Fiber-reinforced polymer composites for construction—State-of-the-art review, *Journal of Composites for Construction*, 6, 2, 73–87.
26. Bakli, D. (1986) Surface phenomena at adhesion and frictional interactions. Moscow, *Mashinostroenie*, 423.
27. Bank, L.C., Mosallam, A.S. (1990) Creep and failure of a full-size fiber reinforced plastic pultruded frame, Composite Material Technology, D. Hui and T. J. Kozik (Eds.), American Society of Mechanical Engineers.
28. Bank, L.C., Gentry, T.R. (1995) Accelerated test methods to determine the long-term behavior of friction composite structures: Environmental effects, *J. Reinforced Plastics and Composites*, 14, 558–587.
29. Bank, L.C., Gentry, T.R. (2001) Development of a pultruded composite material highway guardrail, composites, Part A, *Applied Science and Manufacturing*, 32, 1329–1338.
30. Bank, L.C. (1989) Flexural and shear moduli of full-section fiber reinforced plastic (FRP) pultruded beams, *ASTM Journal of Testing and Evaluation*, 17, 1, 40–45.
31. Bank, L.C., Gentry, T.R., Nuss, K.H., Hurd, S.S., Lamanna, A., Duich, S., Oh, B. (2000) Construction of a Pultruded Composite Structure: A Case Study, *ASCE Journal of Composites for Construction*, 4, 3, 112–119.
32. Bank, L.C., Mosallam, A.S., McCoy, G.T. (1994) Design and performance of fixtures for pultruded frame structures, *J. Reinforced Plastics and Composites*, 13, 3, 199–212.
33. Bank, L.C., Yin, J., Nadipelli, M. (1995) Local buckling of pultruded beams—nonlinearity, anisotropy and inhomogeneity, *Construction and Building Materials*, 9, 6, 325–331.
34. Bank, L.C., Yin, J., Moore, L.E., Evans, D., Allison, R. (1996) Experimental and numerical evaluation of beam-to-column connections for pultruded structures, *J. Reinforced Plastics and Composites*, Vol. 15, 10, 1052–1067.
35. Barbero, E.J. (1999) *Introduction to Composite Materials Design*, Taylor & Francis, Philadelphia, PA.
36. Barbero, E.J. (1996) Prediction of buckling-mode interaction in composite columns, *Mechanics of Composite Materials and Structures*, 7, 269–284.
37. Barbero, E.J. (1991) Pultruded structural shapes: from the constituents to the structural behavior, *SAMPE Journal*, 27, 25–30.
38. Barbero, E.J., GangaRao, H.V.S. (1992) Structural applications of composites in infrastructure, part II, *SAMPE Journal*, 28, 1, 9–16.
39. Barbero, E.J., GangaRao, H.V.S. (1991) Structural applications of composites in infrastructure, part I, *SAMPE Journal*, 27, 6, 9–16.
40. Barbero, E.J., Raftoyiannis, I. (1993) Local Buckling of FRP beams and columns, *ASCE J. of Materials in Civil Engineering*, 5, 3, 339–355.
41. Barbero, E.J., Raftoyiannis, I. (1993) Buckling of Pultruded Composite Columns, *Composite Structures*, 24, 139–147.
42. Barbero, E.J., Raftoyiannis, I. (1994) Lateral and distortional buckling of pultruded i-beams, *Composite Structures*, 27, 261–268.
43. Barbero, E.J., Tomblin, J. (1996) A damage mechanics model for compression strength of composites, *J. Solid Structures*, 33, 29, 4379–4393.

44. Barbero, E.J., Turk, M. (2000) Experimental investigation of beam-column behavior of pultruded structural shapes, *J. Reinforced Plastic and Composites*, 19, 2, 249–265.
45. Barbero, E.J., Makkapati, S., Tomblin, J.S. (1999) Experimental determination of compressive strength of pultruded structural shapes, *Composite Science and Technology*, 59, 2047–2054.
46. Barbero, E.J. (1991) Pultruded structural shapes. Constituents to the structural behavior, *SAMPE Journal*, 27, 1, 25–30.
47. Barbero E.J., Reddy J.N., Teply J.L. (1990) A General Two-Dimensional Theory of Laminated Cylindrical Shells, *AIAA Journal*, 28, 3, 544–553.
48. Barpanda, D., Mantena, P.R. (1994) Effects of time and temperature on the dynamic mechanical properties of pultruded glass-epoxy composites, *Journal of Materials Education*, 16, 1, 407–433.
49. Barpanda, D., Mantena, P.R. (1996) Dynamic mechanical analysis of pultruded glass-graphite epoxy composites at elevated temperatures, *Journal of Reinforced Plastics and Composites*, 15, 5, 497–532.
50. Bear, J., Bachmat, Y. (1991) *Introduction to Modeling of Transport Phenomena in Porous Media*. Kluwer Academic Publishers, London, 553.
51. Beaumont, P.W.R., Sekine, H. (2000) Physical modeling of engineering problems of composites and structures, *Applied Composite Materials*, 7, 1, 13–37.
52. Bhatnagar, A., Lakkad, S.C., Ramesh, C.K. (1981) Creep in high glass content unidirectional and bidirectional GRP Laminated, *Fiber Science and Technology*, 15, 13–25.
53. Binienda, W.K., Wang, A.S.D., Delale, F. (1991) Analysis of bent crack in unidirectional fiber-reinforced composites, *Int. J. Fracture*, 47, 1–24.
54. Blagonadezhin, V.L., Barantsov, A.N., Murzakhanov, R.H. (1987) *Mechanics of Composite Materials*, 5, 859–877.
55. Bockarev, S.V., Gimervert, D.A. (1988) Filtration of polymer binder at winding of composite articles, *Mechanics of Composite Materials*, 6, 116–120.
56. Boisse, P., Borr, M., Buet, K., Cherouat, A. (1997) Finite element simulations of textile composite forming including the biaxial fabric behavior, *Composites, Part B, Engineering*, 28, 4, 453–464.
57. Boisse, P., Buet, K., Gasser, A., Launay, J. (2001) Meso/macro-mechanical behavior of textile reinforcements for thin composites, *Composites Science and Technology*, 61, 3, 395–401.
58. Brown, R. (1982) Competition Chances for growth for epoxy adhesive markets, *Adhesives Eqs.*, 25, 2, 21–25.
59. Buet, G.K., Boisse, P. (2001) Experimental analysis and modeling of biaxial mechanical behavior of woven composite reinforcements. *Exp. Mech.*, 41, 3, 260–269.
60. Bushuev, Y.G., Persin, M.I., Sokolov, V.A. (1994) Carbon-carbon composite materials, Moscow, *Metallurgiya*, 128.
61. Byun, J.H., Whitney, T.J., Du G.W., Chou, T.W. (1991) Analytical characterization of two-step braided composites, *J. Composite Materials* 25, 1599–1618.
62. Cairns, D.S., Lagace, P.A. (1992) Consistent engineering methodology for the treatment of impact in composite materials, *Journal of Reinforced Plastics and Composites*, 11, 4, 395–412.
63. Calich, A.E., Chalykh, A.E. et al. (1998) Diagrams of Phase *State of Polymer Systems*, Moscow-Yanus-Kiev, 178.
64. Canfield, A.R. (1985) Braided carbon-carbon nozzle development. *AIAA Papers*, 1096, 23–32.
65. Cherepanov, G.P. (1983) *Failure Mechanics of Composite Materials*, Moscow, Nauka, 296.
66. Chachad, Y.R., Roux, J.A., Vaughan, J.G. (1996) Thermal investigation of pultruded graphite composites for various processing conditions and thickness, *Engineering Plastics*, 9, 2, 91–108.
67. Chachad, Y.R., Roux, J.A., Vaughan, J.G., Arafat, E.S. (1996) Effect of pull speed on die wall temperatures for flat composites of various sizes, *Journal of Reinforced Plastics and Composites*, 15, 7, 718–739.

68. Chachad, Y.R., Roux, J.A., Vaughan, J.G., Arafat, E.S. (1996) Thermal model for three-dimensional irregular shaped pultruded fiberglass composites, *Journal of Composite Materials*, 30, 6, 692–721.
69. Chachad, Y.R., Roux, J.A., Vaughan, J.G., Arafat, E. (1995) Three-dimensional characterization of pultruded fiberglass-epoxy composite materials, *Journal of Reinforced Plastics and Composites*, 14, 5, 495–512.
70. Chachad, Y.R., Roux, J.A, Vaughan, J.G., Arafat, E.S. (1996) Manufacturing model for three-dimensional irregular pultruded graphite/epoxy composites, *Composites, Part A*, 27A, 3, 201–210.
71. Chachad, Y.R., Roux J.A., Vaughan, J.G., Arafat, E.S. (1996) Effect of pull speed on die wall temperatures for flat composites of various sizes, *Journal of Reinforced Plastics and Composites*, 15, 7, 718–739.
72. Chandra, R., Stemple, A.D., Chopra, I. (1990) Thin-walled composite beams under bending, torsional, and extensional loads, *AIAA Journal of Aircraft*, 27, 7, 619–626.
73. Chaudhari, R. A., Garala, H. J. (1995) Analytical/experimental evaluation of hybrid carbon/glass/epoxy thick-section composites under compression, *Compos. Mater.*, 29, 13, 1695–1718.
74. Chiu, C.H., Lu, C.K., Wu, C.M., (1997) Crushing characteristics of 3D braided composite square tubes. *J. Composite Materials*, 31, 22, 2309–2327.
75. Chou, S., Chen, H.-E. (1995) The weaving methods of three-dimensional fabrics of advanced composite materials, *Composite Structures*, 33, 3, 159–172.
76. Cochran, R., Carl Matson, S. T., Wong, D. (1997) Advanced composite processes for aerospace applications, *42nd International SAMPE Symposium*, 42, May 4–8, 635–640.
77. Cowling, M.J., Hashim, S.A. (2000) Creep analysis of adhesively bonded connections in grp pipes including the effect of Defects, *Composites, Part A: Applied Science and Manufacturing*, 31A, 6.
78. Cox, B.N., Dadkhah, M.S., Inman, R.V., Morris, W.L. (1992) Mechanisms of compressive failure in 3D composites, *Acta Metal. Mater.*, 40, 3285–3298.
79. Cox, B.N., Dadkhah, M.S., Morris, W.L., Flintoff, J.G. (1994) Failure mechanisms of 3D woven composites in tension, compression, and bending, *Acta Metall. Mater.*, 42, 3967–3984.
80. Crea, F., Porco, G., Zinno R. (1997) Experimental Evaluation of Thermal Effects on the Tensile Mechanical Properties of Pultruded GFRP Rods, *Applied Composite Materials*, 4, 3, 133–143.
81. Crossman, F.W., Warren, W.J., Wang, A.S.D., Law, G.E. (1980) Initiation and growth of transverse cracks and edge delamination in composite laminates, Part 2: Experimental correlation, *J. Comp. Mater.*, 14, 88–108.
82. Dadkhah, M.S., Flintoff, J.G., Kniveton, T., Coxm B.N. (1995) Simple models for triaxially braided composites, *Composites*, 26, 91–102.
83. Kim, D.-H., Lee, W.I., Klaus F. (2001) A model for a thermoplastic pultrusion process using commingled yarns, *Composites Science and Technology*, 61, 8, 1065–1077.
84. Dagli, L., Remond, Y. (2002) Identification of the non-linear behaviour a 4d carbon–carbon material designed for aeronautic application, *Applied Composite Materials*, 9, 1, 1–15.
85. Daniel, M., Ishai, O. (1994) *Engineering Mechanics of Composite Materials*, Oxford University Press, New York.
86. Daniel, M., Tsai, C.L. (1991) Analytical/experimental study of cracking in composite laminates under biaxial loading, *Composites Engineering*, 1, 6, 355–362.
87. Dastin, S. (1983) Aircraft composite materials and structures, *Reinforced Plastics*, 27, 1, 15–19.
88. Davalos, J.F., Qiao, P., Barbero, E.J. (1996) Multiobjective material architecture optimization of pultruded FRP I-beams, *Composite Structures*, 35, 3, 271–281.
89. Davalos, J.F., Qiao, P., Salim, H.A. (1997) Flexural-torsional buckling of pultruded fiber reinforced plastic composite I-beams: Experimental and analytical evaluations, *Composite Structures*, 38, 1–4, 8, 241–250.

90. Davalos, J.F., Salim, H.A., Qiao, P., Lopez-Anido, R., Barbero, E.J. (1996) Analysis and design of pultruded FRP shapes under bending, *Composites, Part B, Engineering Journal*, 27B, 295–305.
91. Du, G.W., Ko, F.K. (1996) Analysis of multiaxial warp-knit preforms for composite reinforcement, *Composites Science and Technology*, 56, 253–260.
92. Du, G.W., Ko, F.K. (1993) Unit cell geometry of 3-d braided structures, *Journal of Reinforced Plastics and Composites*, 12, 7, 752–768.
93. Eckold, G.C. (1994) Design and manufacture of composite structures, woodhead, 400.
94. Edwards, P. (1989) New in materials machinery widening application area for pultrusion, *Modern Plastics International*, 19, 6, 30–35.
95. Egorov, L.A., Tyurin, A.A., Mitkevich, A.B. (1987) Squeezing model of binder at composite winding stage, *Mechanics of Composite Materials*, 6, 1127–1129.
96. Esakov A.A., Astrakhantsev, A.L., Kryzhanovsky V.B. (1981) Setting peculiarities of glass-plastic profiles of large cross-sections, *Electrotechnical Industry*, 7, 132, 1–7.
97. Fabeny, B., Curtin, W.A. (1996). Damage-enhanced creep and rupture in fiber-reinforced composites, *Acta Materialia*, 44, 3439–3451.
98. Farris, K.J., Lee, G. (1983) Determination of time independent component of moduls during of thermosetting systems, *Polymer Engineering and Science*, 28, 10, 586–590.
99. Fitzer, E., Manocha, L.M. (1998) *Carbon reinforcements and Carbon / Carbon Composites*, Springer Berlin-Heidelberg, p. 192.
100. Fleck, N.A, Jelf, M., Curtis, PT. (1995) Compressive failure of laminated and woven composites. *J. Composites: Technology and Research*, 17, 3, 212–220.
101. Found, M.S., Holden, G.J., Swamy, R.N. (1997) Static indentation and impact behavior of GRP pultruded sections, *Composite Structures*, 39, 3–4, 223–228.
102. Franke, L., Meyer, H.J. (1992) Predicting the tensile strength and creep-rupture behavior of pultruded glass-reinforced polymer rods, *Journal of Materials Science*, 27, 4899–4908.
103. Freger, D.G. (1999) Development of composite materials for rod structures, *Processing of Materials*, Lugansk, pp. 38–42.
104. Freger, D.G. (2000) *Resource Saving Technological Processes of Manufacture and Pressure Processing of Materials in Mechanical Engineering*, Lugansk, 219–224.
105. Freger, D.G. (1998) Research of technological process of continuous forming of reinforced composites anchor, *Scientific Works of East Ukrainian University*. Lugansk, 72–76.
106. Freger, D.G. (2000) Molding technology design elements of aeroplanes using spatially reinforced fillers, *Proc. Int. Conf.* Composite materials in industry (SLAVPOLIKOM—2000), Kiev, Alkon, 126–127.
107. Freger, G.E. (1982) Stress-strain state investigation of spiral-reinforced materials under shear, *J. Strength Problems*, 11, 116–119.
108. Freger, G.E. (1983) Estimation of elastic characteristics of spiral-reinforced materials with allowance of anisotropy of interlayers, in *Mechanics of Reinforced Plastics*, Riga Polytechn. Inst., Riga, 51–60.
109. Freger, G.E. (1983) Investigation of stress state of highly filled spiral-reinforced composites, Skudra, A.M. (Ed) *Mechanics of Reinforced Plastics*, Riga Polytechn. Inst., Riga, 61–68.
110. Freger, G.E. (1983) Stress-strain state of spiral-reinforced composites at transversal loading, *J. Mech. of Comp. Mater.*, 6, 989–995.
111. Freger, G.E. (1984) Investigation of composite materials based on spiral-reinforced fillers, *J. Mech. of Comp. Mater.*, 3, 412–416.
112. Freger, G.E., Kestelman, V.N. (1990) Auswahl technologischer Parameter zur Herstellung von spiralförming anisotropen Hybridfüllstoffen, *Plaste und Kautschuk*, 37, 10, 349–351.
113. Freger, G.E., Kestelman, V.N., Karvasarskaja, N.Aaxi. (1989) Symmetrical and Thermal Loading of Anisotropic Thick Helical Tubes. *Conf. Fiber Reinf. Pol. Blend.*, 27–28 September.
114. Freger, G.E., Bakst, E.E., Freger, D.G. (1996) *Technologia wykonywania zbrojonych ukladow silowych zlizonej konstrukcji*, Seminarium pt: Wybrane technologie obrobki plastycznej, Poznan, maja 20–21, 75–78.

115. Freger, G.E., Karvasarskaja, N.A. (1993) Calculation and Rational Design of Rods from Composite Materials, *Text book, Kiev, 127.*
116. Freger, G.E., Karvasarskaja, N.A. (1990) Calculation and Optimum Design of Composite Elements of Rod Structures, *Mechanics of Composite Materials*, 3, 501–507.
117. Fuwa, M., Bunsell, A.R., Harris, B. (1975) Tensile failure machanisms in carbon fibre reinforced plastics, *J. Mater. Sci.*, 10, 12, 2062–2070.
118. Gadam, S.U.K., Roux, J.A., McCarty, T.A., Vaughan, J.G. (2000) Pressure rise inside a cylindrical pultrusion die for graphite/epoxy composites, *Polymers and Polymer Composites*, 8, 4, 231–244.
119. Gadam, S.U.K., Roux, J.A., McCarty, T.A., Vaughan, J.G. (2000) The impact of pultrusion processing parameters on resin pressure rise inside a tapered cylindrical die for glass-fibre/epoxy composites, *Composite Science and Technology*, 60, 945–958.
120. Gaidachuk, V.E., Grechka, V.E., Kobrin, V.N. (1986) *Design Principles of Manufacturing Aeroplane Parts from Composite Materials*, Kharkov, Kharkov Aviation Institute, 95.
121. Gaidachuk, V.E., Kiva, D.S., Kobrin, V.N., Rjabov, V.I. (1998) Design and *Manufacturing of Aviation Structures from Composites*, Kharkov, Kharkov Aviation Institute, 128.
122. Gaudenzi, P. (1997) On delamination buckling of composite laminates under compressive loading, *Composite Structures*, 39, 1–2, 10, 21–30.
123. Gause, L.V., Alper, J.M. (1987) Structural properties of braided graphite/epoxy composites, Composites Technology and Research, 9, 4,141–150.
124. Gerdeen, J.C., Snyder, V.W., Viegelahn, G.L., Parker, J. (1979) *Design Criteria for Roof Bolting Plans Using Fully Grouted Non-tensioned Bolts to Reinforce Bedded Mine Roof.* U.S. Department of the Interior, Bureau of Mines, OFR 46–80, vols. III-IV.
125. Gerrard, C. (1983) Rock bolting in theory, *Proceedings of the International Symposium on Rock Bolting*, Abisko, Sweden, 3–29.
126. Gibson, A.G. (1992) Modification of Darcy's law to model mould interface effects in composites processing, *Composites Manufacturing*, 3, 2, 113–118.
127. Gillham, J.K., Enna, J.B. (1983) Time-temperature-transformation (TTT) cure modeling the cure behavior of thermoplast. *Journal of Applied Polymer and Sciencs*, 28, 8, 2567–2591.
128. Gommers, B., Verpoest, I., Van Houtte, P. (1996) Modeling the elastic properties of knitted-fabric-reinforced composites, *Composites Science and Technology*, 56, 6, 685–694.
129. Gommers, B., Verpoest, I., Van Houtte, P. (1998) Analysis of knitted fabric reinforced composites. Part I: Fiber orientation distribution, *Composites* 29A, 1579–1588.
130. Gommers, B., Verpoest, I., Van Houtte, P. (1998) Analysis of knitted fabric reinforced composites. Part II: Stiffness and strength, *Composites*, 29A, 1589–1601.
131. Gorthala, R., Roux, J.A., Vaughan, J.G. (1994) A model to predict resin pressure/back flow in the tapered inlet of a pultrusion die, *Journal of Reinforced Plastics and Composites*, 13, 6, 484–497.
132. Gorthala, R., Roux, J.A., Vaughan, J.G., Donti, R.P. (1994) Comparison of processing parameters for pultruded graphite/epoxy and fiberglass/epoxy: A heat transfer and curing model, *J. Reinforced Plastics and Composites*, 13, 4, 288–300.
133. Gorthala, R., Roux, J.A., Vaughan, J.G. (1994) Resin flow, cure and heat transfer analysis for pultrusion process, *J. Composite Materials*, 28, 6, 486–506.
134. Gowayed, Y., Pastore, C. (1992) Analytical techniques for the prediction of Elastic Properties of Textile Reinforced Composites, *Mechanics of Composite Materials*, 5, 579–596.
135. Gowayed, Y., Wu, J., Barowski, L., Westphall, W. (1996) Mapping of preform architecture for textile reinforced composite products, Composites Part A, *Applied Science and Manufacturing*, 27A, 1023–1032.
136. Gowayed, Y., Zou, W., Gross, S. (2000) An analytical approach to evaluate the coefficients of thermal expansion of textile composite materials, *Polymer Composites*, 21, 5, 814–820.
137. Grabandt, O., Andrews, M.C., Yound, R.J., Mahy, J., Scaap, A.A. (1998) Compressive behavior of aramid fiber-reinforced pultruded rods, *J.Compos. Mater.*, 32, 9, 893–908.

138. Grayson, R.L., Layne, L.A., Althouse, R.C., Klishis, M.J. (1992) Risk indices for roof bolter injuries using microanalysis, *Mining Engineering*, 44, 2, 164–166.
139. Grimes, G.C. (1993) Tape Composite Material Allowables Application in Airframe Design/Analysis, *Composites Engineering*, 3, 7–8, 777–804.
140. Grove, S.M. (1999) Optimum fiber orientation in filament wound structures, *Journal of Materials Science Letters*, 18, 5, 1203.
141. Grushetsky, I.V., Dmitrienko, I.P., Ermolenko, A.F. (1986) Fracture of composite structures, Tamuzh V.P., Protasov V.D. (Eds.) Riga, *Zinatne*, 264.
142. Gudmundson, P., Alpman, J. (2000) Initiation and growth criteria for transverse matrix cracks in composite laminates, *Composites Science and Technology*, 60, 2, 185–195.
143. Hades, I., Vater, B., Darby, M.I., Woctenhalm, G.N. (1969) Residual stresses and the optimum cure cycle ferman epoxy resin, *Mater. Sci.*, 24, 5, 1984–1990.
144. Halm, H.T. (1976) Residual stress in polimer matrix composite laminates, *J. Compos. Mater*, 10, 4, 266–278.
145. Halpin, J.C., Jerine, K., Whitney, J.M. (1971) The laminate analogy for 2 and 3 dimensional composite materials, *J. Compos. Mater.*, 1, 36–49.
146. Hamada, H., Kameo, K., Sakaguchi, M., Saito, H., Iwamoto, M. (2000) Energy-absorption properties of braided composite rods, Composites Science and Technology, 60, 5, 723–729.
147. Hamada, H., Nakatani, T., Nakai, A., Kameo, K. (1999) The crushing performance of a braided I-beam, *Composites Science and Technology*, 59, 12, 1881–1890.
148. Han, C.D., Lee, D.C., Chin, H.B. (1986) Development of a mathematical model for the pultrusion process, *Polymer Engineering and Science*, 26, 6, 393–404.
149. Han, C.D., Lem, K.-W. (1983) Chemorheology of thermosetting resins: II chemorheology and curing kinetics of unsaturated polyester resins, *Journal of Applied Polymer and Sciences*, 28, 10, 3155–3183.
150. Handbook of Composites (1989) vol. 2: Structures and Design. C.T. Herakovich (USA) and Yu. M. Tarnopolskii (Latvia), (Eds.) Amsterdam, the Netherlands, 684.
151. Harte, A.-M., Fleck, N.A. (1999) On the mechanics of braided composites in tension, *European Journal of Mechanics*, 19, 2, 259–276.
152. Hatiashvily, G.M. (1983) AlmansiMichel's problems for homogeneous and composite bodies, Part 1, *Tbilisi, Metsniereba*, 236.
153. Herakovich, C.T. (1998) Mechanics of Fibrous Composites, John Wiley, New York.
154. Hitriov, V.V., Katorzhnov, Yu.I. (1985) Technological means of raising bearing capacity of compressible composite rods, *Mechanics of Composite Materials*, 2, 316–322.
155. Hollaway, L. (Ed.) (1994) *Handbook of Polymer Composites for Engineers*, Woodhead Publishing, Cambridge, England.
156. Hollaway, L., Romhi, A., Gunn, M. (1990) Optimization of adhesive bonded composite tubular sections, *Compos. Struct.*, 1, 125–170.
157. Holmes, M., Rahman, T.A. (1980) Creep behavior of glass reinforced plastic box beams, *Composites*, 4, 79–85.
158. Hou, T.H., Bai, J.M. (1988) A Semi-empirical approach for the chemoviscosity modeling of reactive resin system. *SAMPE Journal*, 24, 6, 43–51.
159. Hsiao, H.-M., Daniel, I.M., Wooh, S.-C. (1995) Fabrication Methods for Unidirectional and Crossply Composites with Fiber Waviness, *J. Advanced Materials*, 26, 2, 19–26.
160. Hsiao, H.-M., Wooh, S.-C., Daniel, I.M. (1995) a new compression test method for thick composites, *J. Composite Materials*, 29, 13, 1789–1806.
161. Hu, J., Jiang, Y., Ko, F.K. (1998) Modeling of uniaxial tensile properties of multiaxial warp knitted fabrics, *Textile Research Journal*, 68, 11.
162. Huysmans, G., Verpoest, I., Van Houtte, P. (2001) A damage model for knitted fabric composites, *Composites A*, 32, 10, 1465–1475.
163. Huysmans, G., Verpoest, I., Van Houtte, P. (1998) A polyinclusion approach for the elastic modeling of knitted fabric composites, *Acta Mater.*, 46, 9, 3003–3013.
164. Iddon, K.J., Blundell, C. (1980) The pultrusion process and its application, *Symp. Fabrication Techniques for Advanced Reinforced Plastics*, Univ. of Salford, April 22-23, TSB4, 58–68.

165. Ifju, P.G. (1995) Shear Testing of Textile Composite Materials, *Journal of Composites Technology and Research*, 17, 3.
166. Incardona, S., Migliaresi, C., Wagner, H.D., Gilbert, A.H., Marom, G. (1993) The mechanical role of the fibre-matrix transcrystalline interphase in carbon fiber reinforced, *J. Polymer Microcomposites, Composite Science and Technology*, 47, 1, 43–50.
167. Irion, M.N., Adams, D.F. (1981) Compression Creep Testing of Unidirectional Composite Materials, *Composites*, 4, 117–123.
168. Ishikawa, T., Chou, T.W. (1982) Stiffness and strength behavior of woven fabric composites, *J. Mater Sci.*, 17, 3211–3220.
169. Ito, M., Chou, T.W. (1998) An analytical and experimental study of strength and failure behavior of plain weave composites, *J. Composite Materials*, 32, 1, 2–30.
170. Iyengar, N., Curtin, W.A. (1997) Time-dependent failure in fiber-reinforced composites by fiber degradation, *Acta Materialia*, 45, 4, 1489–1507.
171. Yoo J.-S,, Lee S.-E., Kim C.-G. (2001) Prediction of mechanical behavior of spatially reinforced composites for kick motor nozzle, *Composite Structures*, 54, 57–65.
172. Jirovec, P. (1978) Wechselwirkung zwischen anker und gebirge, *Rock Mechanics*, 7, 139–155.
173. Byun J.-H. (2000) The analytical characterization of 2D braided textile composites, *Composites Science and Technology*, 60, 5, 705–716.
174. Juan, F.G. (1992) Composite laminated shells under internal pressure, *AIAA Journal*, 6, 1669–1672.
175. Judawisastra, H., Ivens, J., Verpoest, I. (1998) The fatigue behavior and damage development of 3D woven sandwich composites, *Composite Structures*, 43, 35–45.
176. Kalamkarov, A.L., Fitzgerald, S.B., MacDonald, D.O., Georgiades, A.V. (2000) The mechanical performance of pultruded composite rods with embedded fiber-optic sensors, *Composites Science and Technology*, 60, 8, 1161–1169.
177. Karabin, G.J., Debevic, W.L. (1976) *Comparative Evaluation of Conventional and Resin Bolting System*, MESA IR, p. 1033.
178. Karbhari, V.M., Haller, J.E. (1998) Rate and architecture effects on progressive crush of braided tubes, *Composite Structures*, 43, 2, 93–108.
179. Karmis, M., Agioutantis, Z., Stiakakis, C., Stiakakis, N. (2001) An interactive roof bolting selection and performance system, *4th International Symposium for Roof Bolting in Mining*, 6–7 June, Aachen, Germany.
180. Karpov, Y.S. (1997) Edge effects in composite structures and means of their consideration at designing. *Proc. Kharkov Aviation Inst.*, HAI, 245–250.
181. Katsube, N. (1995) Estimation of effective elastic module for composites, *J. Solids and Struct.*, 32, 1, 79–88.
182. Kaw, A.K. (1997) *Mechanics of Composites Materials*, CRC Press, NY.
183. Kendall, K. (1976b) Interfacial cracking of a composite. Part 2. Bending, *J. of Mater. Sci.*, 11, 1263–1266.
184. Kessler, A., Bledzki, A. (2000) Influence of the fiber/matrix-interphase on the post-impact properties of glass/epoxy-laminates, *Advanced Composite Materials*, 9, 2, 109–118.
185. Kestelman, V.N., Freger, G.E., Kac, M.L. (1989) Spannungzustand von Verbundkörpern mit gekrümmt-anisotropen Zwischenschichten, *Plaste und Kautschuk*, 36, 6, 208–210.
186. Khair, A. (1983) Physical and analytical modeling of the behavior of truss bolted mine roofs, *Proc. the Int. Symp. on Rock Bolting*, Abisko, Sweden, 125–141.
187. Ko, F., Malkan, S.R. (1989) Effect of fiber reinforcement geometry on single-shear and fracture behavior of three-dimensionally braided glass epoxy composite pins, *Journal of Composite Materials*, 23, 798–818.
188. Ko, F.K. (1989) Three Dimensional Fabrics for Composites, in *Textile Structural Composites*, T.W. Chou and F. K. Ko, (Eds.), Elsevier.
189. Ko, F.K (1994) *Advanced Braiding*, Proceedings, Texcomp-2, Leuven, Belgium, May 17–19.
190. Ko, F.K., Hartman, D. (1986) Impact behavior of 2D and 3D glass/epoxy composites, *SAMPE Journal*, 22, 4, 26–30.

191. Ko, F.K., Van Vuure, A. (1999) Textile Preforming for Complex Shape Structural Composites, *SAMPE Journal*, 35, 3.
192. Ko, F.K., Scardino, F.L. (1982) Triaxial Fabrics—tensile, shear and burst behaviors, *Textile Research Journal*, 51, 2.
193. Ko, F.K., Krauland, K., Scardino, F. (1983) Weft insertion warp knit for industrial applications: hybrid composites, *Journal of Industrial Fabrics*, 1, 3, 26.
194. Kolosov, A.E. (1987) Impregnation of fibrous fillers by a binder, *Mechanics of Composite Materials*, 5, 878–886.
195. Kolosov, A.E., Karimov, A.A., Khodin, V.G. (1988) Impregnation of fibrous fillers by a polymer binder, *Mechanics of Composite Materials*, 4, 651–659.
196. Korotkov, S.V., Zhovner, B.A. (1991) Investigations by DSC method of antistick additives effect on curing time of epoxy binders during pultrusion, *Plastic Masses*, 5.
197. Kostar, T.D., Chou, T.W. (1994) Microstructural design of advanced multistep three-dimensional braided preforms, *J. of Composite Materials* 28, 13, 1180–1201.
198. Kowsika, M., Mantena, P.R. (1999) Static and Low Velocity Impact Response Characterizations of Pultruded Hybrid Glass/Graphite Epoxy Composite Beams, *Journal of Thermoplastic Composite Materials* 12, 3, 121–132.
199. Kowsika, M., Mantena, P.R. (1996) Optimal Pultrusion Process Conditions for Improving the Dynamic Properties of Graphite-Epoxy Composite Beams, *Materials Evaluation*, 54, 3, 386–392.
200. Kozlov, V.P. (1960) *Sewing and Footwear Needles*, Moscow, Rostekhizdat, 160.
201. Kragelskii, I.V., Dobychin, M.N., Kombalov, V.S. (1977) *Fundamentals of Friction and Wear Calculations*, Moscow, Mashinostroenie, 525.
202. Krishnan, B., Alluri, S., Nidumolu, P., Mantena, P.R., Vaughan, J.G., Kowsika, M. (1995) Ultrasonic and Vibration Methods for the Characterization of Pultruded Composites, *Composites Engineering*, 5, 12, 1433–1451.
203. Krutchkoff, L. (1980) Pultrusion. Part 1. Process converts materials into finished shapes continuously, *Plastics Design and Processing*, 20, 7, 34–38.
204. Kumar, S.S., Mantena, P.R. (1996) Dynamic and static torsional characterization of pultruded hybrid cylindrical composite rods, *Journal of Composite Materials*, 30, 8, 918–932.
205. Lackey, E., Vaughan, J.G. (1994) An analysis of factors affecting pull force for the pultrusion of graphite/epoxy composites, *Journal of Reinforced Plastics and Composites*, 13, 3, 188–198.
206. Lackey, E., Theobald, D., Vaughan, J.G. (1994) Effect of fiber/resin wet-out on pultruded composites, 39th International SAMPE Symposium, 39, April 11–14, 183–193.
207. Lackey, E., Vaughan, J.G., Roux, J.A. (1997) Experimental development and evaluation of a resin injection system for pultrusion, *Journal of Advanced Materials*, 10, 30–37.
208. Lackey, E., Vaughan, J.G. (1997) Evaluation of the quality and consistency of open-bath wet-out techniques for pultruded composites. *J. Proceedings of Composites*, Int. Composites EXPO '97, 1, 27–29, 23A:1–10.
209. Lackey, E., Vaughan, J.G. (1993) Effect of Pultrusion Process Parameters on Flexural Strength, *38th International SAMPE Symposium*, 38, May 10–13, 461–470.
210. Lang, T.A. (1972) Rock reinforcement, *Bulletin of the Association of Engineering Geologists*, IX, 3, 215–239.
211. Lang, T.A. (1961) *Theory and Practice of Rock Bolting*, Transactions, American Institute of Mining, Metallurgical and Petroleum Engineers, 220, 33–348.
212. Lee, J.-W., Daniel, I.M. (1990) Progressive transverse cracking of crossply composite laminates, *J. Composite Materials*, 24, 11, 1225–1243.
213. Lei, C., Cai, Y.J., Ko, F.K. (1992) Finite Element Analysis of 3-D Braided Composites, *Advances in Engineering Software*, 14, Elsevier, 187–194.
214. Li, J., Joshi, S. C., Lam, Y.C. (2002) Curing optimization for pultruded composite sections, *Composites Science and Technology*, 62, 3, 457–467.

215. Gan, L.-H., Ye, L., Mai, Y.-W. (1999) Optimum design of cross-sectional profiles of pultruded box beams with high ultimate strength, *Composite Structures*, 45, 4, 279–288.
216. Lykov, N.V. (1972) *Heat and Mass Exchange*, Moscow, Energiya, 560.
217. Lim, T.C., Fujihara, M., Zako, M., Ramakrishna, S., Shang, H.M. (1999) Deep drawing simulation of knitted fabric composites considering geometrical non-linearity, *Science and Engineering of Composite Materials*, 8, 2, 113–122.
218. Lisovskii, A.F. (1984) *Migration of Metal Melts Sintered in Composite Bodies*, Kiev, Navukova dumka, 256.
219. Locke, J.E. (1997) Thermomechanical stress analysis of inhomogeneous antisymmetric cross-ply laminates, *Computers and Structures*, 62, 1, 25–34.
220. Lomov, S.V., Gusakov, A.V., Huysmans, G., Prodromou, A, Verpoest, I. (2000) Textile geometry preprocessor for mesomechanical models of woven composites, *Composites Science and Technology*, 60, 2083–2095.
221. Lomov, S.V., Huysmans, G., Luo, Y., Parnas, R., Prodromou, A., Verpoest, I., Phelan, F.R. (2001) Textile composites: Modeling strategies, *Composites A*, 32, 10, 1379–1394.
222. Lomov, S.V., Huysmans, G., Verpoest, I. (2001) Hierarchy of textile structures and architecture of fabric geometric models, *Textile Research Journal*, 71, 6, 534–543.
223. Lomov, S.V., Verpoest, I. (2000) Compression of woven reinforcements: a mathematical model, *J. Reinforced Plastics and Composites*, 19, 116, 1329–1350.
224. Loubinoux, D. (1989) La pultrusion des thermoplastigues reinforced. Unmoyen dessal des fils impregnes thermoplastigyes, *Ann. Compos*, 1, 2, 157–164.
225. Lozino-Lozinski (1995) *Composite Materials in Aerospace Design*, Zagainov, (Ed) Kluwer Academic Publishers, 464.
226. Luciano, R., Barbero, E.J. (1995) Formulas for the Stiffness of Composites with Periodic Microstructure, *Int. J. of Solids Structures*, 31, 21, 2933–2944.
227. Luciano, R., Barbero, E.J. (1994) Formulas for the stiffness of composites with periodic microstructure, *Int. J. of Solids and structures*, 31, 2933–2944.
228. Ma, C.C., Lee, K.J., Lee, J.D., Hwang, J.S. (1986) The correlation of processing variables for optimizing it Pultrusion Process, *SAMPE Journal*, 22, 5, 42–48.
229. Maddocks, J.R., McManus, H.L. (1996) Prediction of microcracking in composite laminates under thermomechanical loading, *Polymers and Polymer Composites*, 4, 5, 304–314.
230. Maximov, R.D., Plume, E. (2001) Long-term creep of hybrid aramid/glass fiber reinforced plastics, *Mechanics of Composite Materials*, 37, 4, 271–280.
231. Malik, J. (2001) Optimal length of stressed bolts, *4th International Symposium for Roofbolting in Mining*, 6-7 June, Aachen, Germany.
232. Malkin, A.Y., Kulichikhin S.G. (1985) *Rheology of Polymer Formation and Transformations*, Moscow, Khimiya, 240.
233. Mallick, P.K., Newman, S. (1990) *Composite Materials Technology—Processes and Properties*, New York, Munich, Oxford University Press/Hanser Publishers, 400.
234. Mallik, A.K., Gosh, A., Dittrich, G. (1994) *Kinematic Analysis and Synthesis of Mechanism. CRC Press.*
235. Mark, C., Molinda, G.M., Dolinar, D.R. (2001) Analysis of roof bolt systems. Peng SS, Mark C, Khair AW. (Eds.) *Proc. 20th Int. Conf. on Ground Control in Mining*, Morgantown, WV, West Virginia University, 218–225.
236. Mark, C. (2000) Design of roof bolt systems, Mark, C., Dolinar, D.R., Tuchman, R., (Eds.) *New Technology for Coal Mine Roof Support*, Proc. NIOSH Open Industry Briefing, NIOSH IC 9453, 111–132.
237. Markushev, E.A. (1967) Sewing machines, *Structure and Design Foundations*, Kiev, Tekhnika, 392.
238. Martens, P.N. (1995) Ankerausbau im Bergbau—Roofbolting in Mining, *Zweites Internationales Kolloquium an der RWTH Aachen*, 27–28 März Aachener Beiträge zur Rohstofftechnik und -wirtschaft, 3, 1.
239. Martens, P.N., Rattmann, L. (1998) Developments in rockbolting, *Journal of Mines, Metals & Fuels, Mining Industry Annual Review*, 405–411.

240. Martin, J.D. (1983) Pultrusion, plastics products, *Design Handbook, Part B*, Marcel Dekker Inc., New York, 37–74.
241. Martin, J.D. (1979) Pultrusion–New processes, *Plastics Engineering*, 35, 3, 53–57.
242. Masters, J.E., Foye, R.L., Pastore, M., Gowayed, Y.A. (1993) Mechanical properties of triaxially braided composites: Experimental and Analytical Results, *J. of Composites Technology and Research*, 15, 2, 112–122.
243. Maximov, R.D., Plume, E.Z. (1980) Elasticity of a hybrid composite material based on organic and boron fibers, *J. Mech. of Comp. Mater.*, 3, 399–403.
244. Mazumdar, S.K. (2002) Composites manufacturing materials, *Product and Process Engineering*, CRC Press, 407.
245. McGrath, J.E., Grubbs, H., Woodward, M.H., Rogers, M.E., Gungor, A., Joseph, W.A., Mercier, R., Brennan, A. (1994) Development of new polymeric matrices for high performance composites, *Composite Structures*, 27, 7–16.
246. Mehan, M.L., Schadler, L.S. (2000) Micromechanical behavior of short-fiber polymer composites, *Composites Science and Technology*, 60, 7, 1013–1026.
247. Meshii, M., Raghavan, J. (1994) Activation theory for creep of matrix resin and carbon fiber reinforced polymer composites, *J. Mat. Science*, 5078–5084.
248. Meshii, M., Raghavan J. (1998) Creep of Polymer Composites, *Composite Science and Technology*, 57, 1673–1688.
249. Meshii, M., Raghavan, J. (1997) Creep Rupture of Polymer Composites, *Composite Science and Technology*, 57, 4, 375–388.
250. Meshii, M., Raghavan, J. (1995) Prediction of Creep Rupture of Unidirectional Carbon Fiber Reinforced Polymer Composite, *Mat. Sci. & Eng.* A 197, 237–249.
251. Meshii, M., Raghavan, J. (1996) Time-dependent Damage in Carbon Fiber Reinforced Polymer Composites, *Composites A*, 27A, 1223–27.
252. Meyer, R.W. (1985) Handbook of pultrusion technology, New York, Chapman and Hall, 188.
253. Michii, Y.C., McManus, H.L. (1997) Prediction of microcracking distributions in composite laminates using a Monte-Carlo simulation method, *Journal of Reinforced Plastics and Composites*, 16, 13, 1220–1230.
254. Mohajerjasbi, S. (1996) Structure and properties of three-dimensional braided composites, including axial yarns, *AIAA Journal*, 34, 1, 209–211.
255. Mohajerjasbi, S. (1997) Predictions for coefficients of thermal expansion of three-dimensional braided composites, *AIAA Journal*, 35, 1, 141–144.
256. Mohajerjasbi, S. (1998) Fiber Architecture of Three-Dimensional Braided Composites, *AIAA Journal*, 36, 4, 613–617.
257. More composites in commercial transports? (1989) *Aerosp. Eng.*, 9, 12, 19–22.
258. Morton, S. K., Webber, J.P.H. (1994) Optimal design of a composite I-beam, *Composite Structures*, 28, 149–168.
259. Mosallam, A.S., Bank, L.C. (1992) Short term behavior of a pultruded fiber reinforced plastic frame, *ASCE Journal of Structural Engineering*, 118, 7, 1937–1954.
260. Mottram, J.T., Zheng, Y. (1996) State-of-the-art review on the design of beam-to-column connections for pultruded frames, *Composite Structures*, 35, 4, 387–401.
261. Mottram, J.T. (1993) Short- and long-term structural properties of pultruded beam assemblies fabricated using adhesive bonding, *Composite Structures*, 25, 387–395.
262. Mudryi, A.A., Taranenko, I.M. (1999) Edge effects in composite beams, Aviation and cosmic technology and machinery, *Proc. Kharkov Aviation Inst.*, HAI, 20–25.
263. Nagaraj, V., Hota, V.S., GangaRao (1997) Static behavior of pultruded GFRP beams, *Journal of Composites for Construction*, 1, 3, 120–129.
264. Naik, N.K. (1994) *Woven Fabric Composites*, Technomic Publishing Company.
265. Naik, R.A., Ifju, P.G., Masters, J.E. (1994) Effect of fiber architecture parameters on deformation fields and elastic moduli of 2D braided composites, *Journal of Composite Materials*, 28, 7, 656–681.
266. Naik, R.N. (1994) *Analysis of Woven and Braided fabric Reinforced Composites*, NASA CR-19493.
267. Nori, C.V., McCarty, T.A., Mantena, P.R. (1996) Vibration analysis and finite element modelling for determining shear modulus of pultruded hybrid composites, *Composites Engineering, Part B*, 27B, 329–337.

268. Oillham, J.K., Enns, J.B. (1983) Time-temperature transformation (TTT) cure diagrams modeling the cure behavior of thermosets, *J. of Applied Polymer and Science*, 28, 8, 567–2591.
269. Olivier, Siron, O., Pailhes, J., Lamon, J. (1999) Modeling of the stress/strain behavior of a carbon/carbon composite with a 2.5 dimensional fibre architecture under tensile and shear loads at room temperature, *Composites Science and Technology*, 59, 1, 1–12.
270. Ostermayer, H., Scheele (1977) Research on ground anchors in non-cohesive soils. *Revue Francaise de Geotechnique*, 3, 92–97.
271. Paciornik, S., Martinho, F.M., de Mauricio, M.H.P., d'Almeida, J.R.M. (2003) Analysis of the mechanical behavior and characterization of pultruded glass fiber–resin matrix composites, *Composites Science and Technology*, 63, 2, 295–304.
272. Palmer, D.W., Bank, L.C., Gentry, T.R. (1998) Progressive tearing failure of pultruded composite box beams: experiment and simulation, *Composites Science and Technology*, 58, 8, 1353–1359.
273. Pandey, R., Hahn, H.T. (1996) Designing with 4-step braided fabric composites, *Composites Science and Technology*, 56, 6, 623–634.
274. Pandita, S.D., Falconet, D., Verpoest, I. (2002) Impact properties of weft knitted fabric reinforced composites, *Composites Science and Technology*, 62, 7–8, 6, 1113–1123.
275. Panek, L. (1956) Principles of reinforcing bedded mine roof with bolts, U.S. Bureau of Mines RI 5156, 25.
276. Pang, F., Bathgate, R., Wang, C. (1997) Creep and creep rupture of stitched composites, 11th International Conference on Composite Materials, Australian *Composite Structures Society*, Australia, 97–106.
277. Pang, F., Wang, C., Bathgate, R. (1997) Creep response of woven-fibre composites and the effect of stitching, *Journal of Composite Science and Technology*, 56, 91–98.
278. Park, C.H., McManus, H.L. (1996) Thermally induced damage in composite space structures: predictive methodology and experimental correlation, *Composites Science and Technology*, 56, 10, 1209–1219.
279. Parsons, F.W. (1964) Development and testing of an explosive-anchored rock-bolt, *Mining Congress Journal*, 6, 29–32.
280. Pastore, C.M., Masters, J.E., Gowayed, Y.A., Foye, R.L. (1993) Mechanical properties of triaxially braided composites: Experimental and theoretical results, *Journal of Composites Technology and Research*, 15, 2, 112–122.
281. Pastore, C.M., Ko, F.K. (1991) Modeling of textile structural composites-part 1: Processing science model for three-dimensional braiding, *Journal of the Textile Institute*, 81, 4.
282. Pender, E.B., Hosking, A.D., Mattner, R.H. (1963) Grouted rock bolts for permanent support of major underground works, *Journal of the Institute of Engineers*, Australia, 35, 7–8, 129–150.
283. Peng, S.S. (1998) Roof bolting adds stability to weak strata, *CoalAge Magazine*, 12, 32–38.
284. Penn, L.S., Chou, C.T., Wang, A.S.D., Binienda, W.K. (1989) The effect of matrix shrinkage on damage accumulation in composites, *J. Composite Materials*, 23, 570–586.
285. Perov, Y.Y. (1991) Prevention of edge separation of laminated plastics, *Mechanics of Composite Materials*, 3, 468–473.
286. Philips, D., Verpoest, I., Van Raemdonck, J. (1996) 3D-knitted fabrics for sandwich panels, *SAMPE Journal*, 32, 6, 23–31.
287. Piggot, M.R., Harris, B. (1981) Compression strength of hybrid fiber-reinforcement plastics, *J. Mater. Sci.*, 16, 13, 678–693.
288. Plastics of structural purposes (1974) Trostyanskaya, E.B. (Ed), Khimiya, Moscow, 304.
289. Polilov, A.N., Rabotnov, Y.N. (1983) On delamination of composite pipes in the form of Chinese lantern, *Mechanics of Composite Materials*, 3, 548–560.
290. Polilov, A.N., Rabotnov, Y.N. (1983) Propagation of delamination at compression of composites, *Solid Mechanics*, 4, 166–171.

291. Poss, A.L. (1975) Designing with three directional composites, *Mech. Eng.*, 4, 32–37.
292. Prodromou, A.G., Chen, J. (1997) On the relationship between shear angle and wrinkling of textile composite preforms, *Composites Part A*, 28, 491–503.
293. Pultrusion—unlimited potential? (1995) *Adv. Mat. News*, 71, 10–11.
294. Qiao, P., Davalos, J.F., Barbero E.J. (1998) Design optimization of fiber reinforced plastic composite shapes, *J. Composite Materials*, 32, 2, 177–196.
295. Ramani, K., Dai, H. (2002) Design and process for preformed woven, knitted, and braided thermoplastic composite reinforced arrester, *Journal of Composite Materials*, 36 20, 2357–2372.
296. Raper, K., Roux, J.A., McCarty, T.A., Vaughan, J.G. (2000) Inlet contour impact on the pressure rise in a pultrusion die, *Journal of Composite Materials*, 34, 3, 199–217.
297. Raper, K., Roux, J.A., McCarty, T.A., Vaughan, J.G. (1999) Investigation of the pressure behavior in a pultrusion die for graphite/epoxy composites, *Composites, Part A, Applied Science and Manufacturing*, 30, 1123–1132.
298. Raper, K.S., Roux, J.A., Vaughan, J.G., Lackey, E. (1999) Permeability impact on the pressure rise in a pultrusion die, *Journal of Thermophysics and Heat Transfer*, 13, 12, 91–99.
299. Roberts, T.M., Al-Ubaidi, H. (2002) Flexural and torsional properties of pultruded fiber reinforced plastic I-profiles, *Journal of Composites for Construction*, 6, 1, 28–34.
300. Roberts, T.M. (2002) Influence of shear deformation on buckling of pultruded fiber reinforced plastic profiles, *Journal of Composites for Construction*, 6, 4, 241–248.
301. Rosato, D.V. (1997) *Designing with Reinforced Plastics*, Hanser/Gardner, Cincinnati, Ohio.
302. Roubinet, P. (1986) La pultrusion courle, *Composites*, 4, 69–73.
303. Roux, J.A., Vaughan, J.G., Shanku, R., Arafat, E.S., Bruce, J., Johnson, V.R. (1998) Comparison of measurements and modeling for pultrusion of a fiberglass/epoxy i-beam, *Journal of Reinforced Plastics and Composites*, 17, 17, 1557–1579.
304. Ruan, X., Chou, T.-W. (1996) Experimental and theoretical studies of the elastic behavior of knitted-fabric composites, *Composites Science and Technology*, 56, 12, 1391–1403.
305. Rubinstein, A. A. (1998) The fracture analysis of composites by a micromechanical approach, *Composites Science and Technology*, 58, 11, 1785–1792.
306. Sabnis, G. M. (1979) *Handbook of Composite Construction Engineering*, Van Nostrand Reinhold Company, New York.
307. Saito, H., Chirwa, E. C., Inai, R., Hamada, H. (2002) Energy absorption of braiding pultrusion process composite rods, *Composite Structures*, 55, 4, 407–417.
308. Savci, S., Curiskis J. I. and Pailthorpe M. T. (2000) A study of the deformation of weft-knit preforms for advanced composite structures, Part 2: The resultant composite, *Composites Science and Technology*, 60, 10, 1943–1951.
309. Savin, A.G., Bokii M.A., Obukhova N.S. (1988) Chemoviscous model of nonisothermal curing of thermosetting binders, *Mechanics of Composite Materials*, 6, 884–887.
310. Shcerbakov, V.T., Muratov, V.M., Zhovner, B.A., Prosvetova, L.A. (1987) Production of continuous profiles from carbon plastics by pultrusion, *Aviation industry*, 6, 50–51.
311. Schmuck C.H. (1979). Cable bolting at the Homestake mine, *Mining Engineering*, 11, 1677–681.
312. Schmuck, H.K. (1957) Theory and practice of rock bolting, *Colorado School of Mines Quarterly*, 52, 3, 233–263.
313. Scida D., Aboura, Z., Benzeggagh, M.L., Bocherens, E. (1999) A micromechanical model for 3D elasticity and failure of woven-fibre composite materials, *Composites Science and Technology*, 59, 505–517.
314. Scott, D.W., Zureick, Abdul-H. (1998) Compression creep of a pultruded E-glass/vinylester composite, *Composites Science and Technology*, 58, 8, 1361–1369.
315. Scott, J.J. (1992) Roof bolting in mining-USA. *Proceeding, Norsk Jord-og Fjellteknisk Forbuns*, Oslo, Norway, November 27, 13.
316. Shanku, R., Vaughan, J.G., Roux, J.A. (1997) Rheological characterization and cure kinetics of epon 862/w epoxy used in pultrusion, *Advances in Polymer Technology*, 16, 4, 297–311.

317. Sharma, D., McCarty, T.A., Roux, J.A., Vaughan, J.G. (1998) Fluid mechanics analysis of a two-dimensional pultrusion die inlet, *Polymer Engineering*, 38, 10, 1611–1622.
318. Sharma, D., McCarty, T.A., Roux, J.A., Vaughan, J.G. (1998) Investigation of dynamic pressure behavior in a pultrusion die, *Journal of Composite Materials*, 32, 10, 929–950.
319. Sharma, D., McCarty, T.A., Roux, J.A., Vaguhan, J.G. (1998) Pultrusion die pressure response to changes in die inlet geometry, *Polymer Composites*, 19, 2, 180–192.
320. Shivakumar, K.N., Emehel, T.C., Avva, V.S., Sadler, R.L. (1995) Compression strength and failure mechanisms of 3D textile composites, *AIAA*, 95–1159.
321. Shun-FaHwang, Ching-Ping, M. (2001) Failure of delaminated interply hybrid composite plates under compression, *Composites Science and Technology*, 61, 11, 1513–1527.
322. Shirokov, A.P. (1981) *The Theory and Practice of Roof Bolting Application*, Moscow, Nedra, 391.
323. Shirokov, A.P., Lider, V.A., Dzaurov, M.A. (1990) *Roof Bolting*, Ref. Book, Moscow, Nedra, 205.
324. Siron, O., Pailhes, J., Lamon, J. (1999) Modeling of the stress/strain behaviour of a carbon/carbon composite with a 2.5 dimensional fibre architecture under tensile and shear loads at room temperature, *Composites Science and Technology*, 59, 1, 1–12.
325. Skudra, A.M., Bulavs, F.Y. (1982) *Strength of Reinforced Plastics*, Khimiya, Moscow, 233.
326. Skudra, A.M., Bulavs, F.Y., Rotsens, K.A. (1971) *Creep and Static Fatigue of Reinforced Plastics*, Riga, *Zinatne*, 238.
327. Smith, L.V., Swanson, S.R. (1995) Failure of braided composite cylinders under biaxial tension, *J. Compos. Mater*, 29, 6, 766–784.
328. Smith, S.J., Bank, L.C., Gentry, T.R., Nuss, K.H., Hurd, S.H., Duich, S.J., Oh, B. (2000) Analysis and testing of a prototype pultruded composite causeway structure, *Composite Structures*, 49, 2, June, 141–150.
329. Souter, B.J., Robitaille, F., Long, A.C., Rudd, C.D. (1998) Geometric description of woven and warp-knitted industrial reinforcements for the calculation of preform deformation and flow properties, *Proceedings of Automotive Composites Workshop*, 35–43.
330. Steiner, R.L., Cole, J.D., Strong, A.B., Todd, R.H. (1992) Recommendations for composite manufacturing pultrusion process and equipment, *SAMPE Quarterly*, 24, 1, 38–44.
331. Stimpson, B. (1987) Estimating the degree of reinforcement of horizontally bedded roofs by non-tensioned, full-column grouted bolts, *Int. J. Min. and Geol. Eng.*, 5, 273–284.
332. Stimpson, B. (1987) Optimizing the reinforcement effect of full-column, untensioned, grouted bolts in bedded mine roofs, *Int. J. Min. and Geol. Eng.*, 5, 285–298.
333. Stowell, E.A., Lin, T.S. (1967) On mechanical behavior of fiber-reinforced cristalline materials, *J. Mech. and Phys. of Solidas*, 9, 4, 242–246.
334. Sukiasov, V., Wahren, A., Kushnevski, V. (1998) Mathematische modelierung des aushärte—prozesses von dünwandigen konstruktionselementen aus verbundwerkstoffen, *Forschung im Ingenieurwesen*, 64, 129–135.
335. Sukiasov, V.G., Altenbach, H., Kushnevsky, V. (2000) Analyse technologisch bedingter Spannungen und Verzerrungen in dünnwandigen Bauteilen aus verstärkten Kunststoffen, *Forschung in Ingenieurwesen*, Berlin 65 8, 207–216.
336. Sumerak, J.E. (1985) Understanding of pultrusion process variables for the first time, *Plastics Technology*, March, 83–85.
337. Sumerak, J.E., Martin, J.D. (1984) It's time we really understood pultrusion process variables, *Plastics Technology*, February, 13–19.
338. Swanson, S.R., Smith, L.V. (1996) Strength design with 2-D triaxial braid textile composites, *Composites Science and Technology*, 56, 3, 359–365.
339. Tabiei, A., Svenson, A., Hargarve, M., Bank, L. (1998) Impact performance of pultruded beams for highway safety applications, *Composite Structures*, 42, 3, 231–237.

340. Tabiei, A., Svanson, A., Hargrave, M. (1996) Impact behavior of pultruded composite box-beams, *Composite Structures*, 36, 1–2, 10, 155–160.
341. Takeda, N., N., McCartney, L., Ogihara, S. (2000) The application of a ply-refinement technique to the analysis of microscopic deformation in interlaminar-toughened laminates with transverse cracks, *Composites Science and Technology*, 60, 2, 231–240.
342. Tang, Z.X., Postle, R. (2001) Mechanics of three-dimensional braided structures for composite materials. Part II: prediction of the elastic moduli, *Composite Structures*, 51, 4, 451–457.
343. Tang, Z.X., Postle, R. (2000) Mechanics of three-dimensional braided structures for composite materials. Part I. Fabric structure and fiber volume fraction, *Composite Structures*, 49, 4, 451–459.
344. Tarnopolskii, Y.M., Arnautov, A.K., Kulakov, V.L. (1999) Methods of determination of shear properties of textile composites, *Composites Part A: Applied Science and Manufacturing*, 30, 879–885.
345. Tarnopolskii, Y.M., Kincis, T.Y. (1985) *Static Test Methods for Composites*, New York: Van Nostrand Reinhold Co., 301.
346. Tarnopolskii, Y.M., Khitrov, V.V. (1986) Composite rods for truss structures, *Mechanics of Composite Materials*, 2, 258–268.
347. Tarnopolskii, Y.M., Kulakov, V.L., Arnautov, A.K. (2000) Measurements of shear characteristics of textile composites, *Computes & Structures*, 76, 115–123.
348. Tarnopolskii, Y.M., Zhigun, I.G., Polyakov, V.A. (1993) *Spatially Reinforced Composites*, Technomic Publishing Co., Inc., 341.
349. Temam, R. (1981) *Navier-Stoke's Equations*, Moscow.
350. Theocaris, P.S., Paipetis, S.A., Stassinakis, C.A. (1978) Effect of geometry and imperfect in composite systems with limiting shear properties, *Fiber Sci. and Technol.*, 11, 5, 335–352.
351. Thomas, W.F. (1977) The effect of fibre diameter on the strength and stiffness of glass fibre reinforced polymeric resins, *Fibre Reinf. Mater. Des. and Eng. Appl. Proc. Conf.*, London, 57–61.
352. Tirosh, J., Katz, E., Lifschuetz, G. (1979) The role of fibrous reinforcements well bonded or partially of composite materials, *Eng. Fract. Mech.*, 12, 2, 267–277.
353. Tolks, A.M., Repelis, I.A., Chailints, M.P., Kantsevich, V.A. (1982) Solid-woven carcasses for spatial reinforcement, *Mechanics of Composite Materials*, 1, 20–25.
354. Tomashevskii, V.T., Shalygin, V.N., Romanov, D.A., Zinoviev, R.S. (1985) Optimum conditions for efficient reinforcement of composite shells by short-fiber particles, *Mechanics of Composite Materials*, 2, 211 216.
355. Tomblin, J., Barbero, E.J. (1994) Local buckling experiments on FRP columns, *Thin-Walled Structures*, 18, 97–116.
356. Tong, L., Steven, G.P. (1996) A numerical study on stresses in resin transfer moulding lap joints reinforced with transverse stitching, *Composite Structures*, 36, 1–2, 10, 91–100.
357. Tong, L., Mouritz, A.P., Bannister, M. (2002) 3D *Fiber Reinforced Polymer Composites*, Elsevier.
358. Truckenmuller, F., Fritz, H. (1991) Injection molding of long fibers—reinforced thermoplastics, comparison of extruded and pultruded materials with direct addition of roving strands, *Polym. Eng. and Sci.*, 31, 18, 1316–1329.
359. Tsai, C.-L., Daniel, I.M. (1991) The behavior of cracked cross-ply composite laminates under simple shear loading, *Composites Engineering*, 1, 1, 3–11.
360. Turve, G.J. (1996) Effects of load position on the lateral buckling response of pultruded GRP cantilevers—Comparisons between theory and experiment, *Composite Structures*, 35, 1, 33–47.
361. Turvey, G.J. (1997) Analysis of pultruded glass reinforced plastic beams with semi-rigid end connections, *Composite Structures*, 38, 1–4, 8, 3–16.
362. Turvey, G.J. (1998) Torsion tests on pultruded GRP sheet, *Composites Science and Technology*, 58, 8, 1343–1351.
363. Turvey, G.J. (1996) Effects of load position on the lateral buckling of pultruded GRP cantilevers. Correlations between theory and experiment, *Composite Structures*, 35,1, 33–47.

364. Turvey, G.J. (1996) Lateral Buckling Tests on Rectangular Cross-Section Pultruded GRP Cantilever Beams, *Composites Part B: Engineering*, 27B, 1, 35–42.
365. Turvey, G.J., Osman, M.Y. (1996) Elastic and initial flexural failure analysis of unsymmetrically laminated cross-ply strips. *Composites Part B: Engineering*, 27B, 5, 505–518.
366. US Patent 4, 591, 400. Method of forming a fiber reinforced composite of a complex configuration, Fradenburgh E.A., Kiely E.F., Miller G.G., May 5 (1984).
367. US Patent 4, 803, 819. Utility pole and attachments formed by pultrusion of dielectric insulating plastic, such as glass fiber reinforced resin, Kelsey F., November 3 (1986).
368. US Patent 4,842,667. Epoxy resin system and pultrusion process employing same, Thorsted, Jr., October 19 (1987).
369. US Patent 5,556,496. Pultrusion method for making variable cross-section thermoset articles, Sumerak J.E., January 10, (1995).
370. US Patent 6,033,510. Method for fabricating composite structures using pultrusion processing, Farley G.L., September 23 (1997).
371. Ustinova, A.M., Oleinokova, E.V., Lipskaya, V.A. (1983) A quick-setting epoxy binder, *Plastic Masses*, 3, 34–35.
372. Valliappan, M., Roux, J.A., Vaughan, J.G. (1996) Temperature and cure in pultruded composites using multi-step reaction model for resin, *Journal of Reinforced Plastics and Composites*, 15, 3, 295–321.
373. Valliappan, M., Roux, J.A., Vaughan, J.G., Arafat, E.S. (1996) Die and post-die temperature and cure in graphite/epoxy composites, *Composites. Part B*, 27B, 1, 1–9.
374. Van de Velde, K., Kiekens, P. (2001) Thermoplastic pultrusion of natural fibre reinforced composites, *Composite Structures*, 54, 2–3, 355–360.
375. Van Vuure, A.W., Ivens, J., Verpoest I. (2000) Mechanical properties of composite panels based on woven sandwich-fabric preforms, *Composites* A, 31, 671–680.
376. Van Vuure, A.W., Ivens, J., Verpoest, I. (1995) Textiele vezelarchitecturen voor kunststof composieten, *Kunststofen Rubber*, 11, 13–17.
377. Vandeurzen, Ph., Ivens, J., Verpoest, I., (1996) A three-dimensional micromechanical analysis of woven fabric composites—I, *Geometric analysis, Composite Science and Technology*, 56, 1303–1315.
378. Vandeurzen, Ph., Ivens J., Verpoest I. (1997) TexComp: a 3D analysis tool for 2D woven fabric composites, *SAMPE Journal*, 33, 2, 25–33.
379. Vandeurzen, Ph., Ivens, J., Verpoest, I. (1996) Structure-performance analysis of two-dimensional woven fabric composites, *Polymers and Polymer Composites*, 4, 361–367.
380. Vanin, G.A. (1985) *Micromechanics of Composite Materials*, Moscow, Mashinostroenie, 302.
381. Vasiliev, V.V. (1988) *Mechanics Structures from Composite Materials*, Moscow, *Mashinostroenine*, 272.
382. Vasiliev, V.V. (1986) *Polymer Compositions in Mining*, Moscow, Nauka, 255.
383. Vasiliev, V.V., Levachenko, V.I. (1991) *The Technology of Physico-Chemical Reinforcement of Rocks*, Moscow, Nedra, 267.
384. Vaughan, J.G., Dillard T.W., Seal E. (1990) A Characterization of the important parameters for graphite/PEEK pultrusion, *Journal of Thermoplastic Composite Materials*, 3, 2, 131–149.
385. Vaughan, J.G., Gorthala, R, Roux, J.A. (1991) Characterization and optimization of the pultrusion process for structural shapes, *Advanced Composite Materials in Civil Engineering Structures, ASCE*, February, 12–23.
386. Verpoest, I., Dendauw, J. (1992) Mechanical properties of knitted glass fibre/epoxy resin laminates, *Proc. of the 37th International SAMPE-Symposium*, Anaheim (USA), 369–377.
387. Verpoest, I., Wevers, M., Ivens, J., De Meester P. (1990) 3D-fabrics for compression and impact resistant composite sandwich structures, *Proc. of the 35th International SAMPE Congress*, Anaheim (USA), 296–307.
388. Vigo, T.L. (1994) *Textile Processing and Properties*, Elsevier Science, 498.

389. Vinson, J.R., Chou, T.W. (1975) *Composite Materials and Their Use in Structures*, London, XII, 438.
390. Wagner, H.D., Marom G. (1982) On composition parameters for hybrid composite materials, *Composites*, 13, 18–20.
391. Wagner, H.D. (1995) Interfaces in microcomposites and macrocomposites: the issue of thermal residual stresses, *Composite Interfaces*, 2, 5, 321–336.
392. Wagner, H.D., Nairn, J.A., Detassis, M. (1995) Toughness of interfaces from initial fiber-matrix debonding in a single-fiber composite fragmentation test, *Applied Composite Materials*, 2, 2, 107–117.
393. Wang, A. S. D., Crossman, F.W. (1980) Initiation and growth of transverse cracks and edge delamination in composite laminates, Part 1: An Energy Method, *J. Composite materials*, 14, 71–87.
394. Wang, A. S. D. (2000) A Design and analysis methodology for 3D braided structural shapes, *Proc. 15th Annual Technical Meeting*, American Society of Composties, Technomic Publishers, 772–779.
395. Wang, A. S. D., Kumar, A. (1998) Local and global properties in 3-D braided structural shapes, *Proc. 8th Japan-US Conference on Composite Materials*, Baltimore, September, 157–166.
396. Wang, A. S. D. (1999) *Micromechanics Analysis of Composite Materials in Mechanics of Composite Materials and Structures*, Kluwer Academic Publisher, 81–130.
397. Wang, S.D., Crossman, F.W. (1977) Edge effects on thermally induced stresses in composite laminates, *J. Composite Materials*, 11. 300–312.
398. Wang, Y.Q., Wang, A. S. D. (1995) Geometrical mapping of yarn structures due to shape change in 3-D braided composites, *J. Composites Science & Technology*, 54, 359–370.
399. Wang, Y.Q., Wang, A. S. D. (1994) On the topological yarn structure in 3-d rectangular and tubular braided preforms, *J. Composites Science & Technology*, 51, 575–586.
400. Wang, Y. Q., Wang, A. S. D. (1997) Spatial distribution of yarns and mechanical properties in 3D braided tubular composites, *J. Applied Composite Materials*, 4, 121–132.
401. Wang, Y., Gowayed, Y., Kong, X., Li. J., Zhao, D. (1995) Properties and analysis of composites reinforced with e-glass weft knitted fabrics, *ASTM Journal of Composite Technology & Research*, 17, 4, 283–288.
402. Wardle B.L., Lagace P.A. (1998) On the behavior of composite shells under transverse impact and quasi-static loading, *AIAA Journal*, 36, 6, 1998, 1065–1073.
403. Wardle, B.L., Lagace, P.A. (1998) On the use of quasi-static testing to assess impact damage resistance of composite shell structures, *Mechanics of Composite Materials and Structures*, 5, 1, 103–121.
404. Kuo, W.S., Ko, T.H., Lo, T.S. (2002) Failure behavior of three-axis woven carbon/carbon composites under compressive and transverse shear loads, *Composites Science and Technology*, 62, 7–8, 989–999.
405. Wham, B.I., Palazotto, A.N. (1996) An investigation of graphite PEEK composite under compression with a centrally located circular discontinuity, *Composite Structures*, 35, 4, 375–386.
406. Whitecomb, J., Kondaguta, G., Woo, K. (1995) Boundary effects in woven composite, *Journal of Composite Materials*, 29, 4, 508–524.
407. Windsor, C.R. (1997). Rock reinforcement systems, *Int. J. Rock Mech. Min. Sc. & Geomech. Abstr*, 34, 6, 919951.
408. Windsor, C.R., Thompson, A.G. (1993) *Rock reinforcement—technology, testing, design & evaluation. Comprehensive Rock Engineering*, J.A. Hudson (Ed.), Oxford: Pergamon, 4, 451–484.
409. Wisnom, M.R. (1995) The effect of transverse compressive stresses on tensile failure of glass fibre-epoxy, *Composite Structures*, 32, 1–4, 621–626.
410. Wisnom, M.R. (1993) Analysis of shear instability in compression due to fibre waviness, *J. Reinfor. Plast. and Compos.*, 12, 11, 1171–1189.
411. Wood, A.S. (1983) Pultrusion: it's getting to be something extra in BP processing, *Modern Plastics*, 60, 7, 50–53.

412. Wood, J.R., Wagner, H.D., Marom, G.A. (1994) Model for Compressive Fragmentation, *Advanced Composites Letters*, 34, 133–138.
413. Wu, D.L. (1996) Three-cell model and 5D braided structural composite, *Composites Science and Technology*, 56, 3, 225–233.
414. Wu, H.-T., Joseph, B. (1990) Model based and knowledge based control of pultrusion process, *SAMPE J*, 26, 6, 59–70, 7, 50–53.
415. Xiao, L.L., Crouch, I.G., Lam, Y.C. (2000) Simulation of heat transfer and cure in pultrusion with a general-purpose finite element package, *Composites Science and Technology*, 60, 6, 857–864.
416. Xiaoyu, J., Xiangan, K. (1999) Micro-mechanical characteristics of fibre/matrix interfaces in composite materials, *Composites Science and Technology*, 59, 5, 635–642.
417. Kogo Y., Hatta, H., Toyoda, M., Sugibayashi, T. (2002) Application of three-dimensionally reinforced carbon–carbon composites to dovetail joint structures, *Composites Science and Technology*, 62, 16, 2143–2152.
418. Uetsuji Y., Zako, M. (1999) Damage Analysis of Woven Fabric Composite Materials, *JSMS*, 48, 9, 1029–1034.
419. Yau, S.S., Chou, T.W., Ko F.K. (1986) Flexural and axial compressive failures of three-dimensionally braided composite I-beams, *Journal of Composite Materials*, 17, 3, 227–232.
420. Yavin, B., Wagner, H.D. (1993) Micromechanical measurements of interfacial adhesion in e-glass/epoxy composites, *Advanced Composites Letters*, 2, 2, 47–50.
421. Yeow, Y.F., Morris, D.H., Briason, H.P. (1979) The fracture behavior of grafite/epoxy laminates, *Exp. Mech.*, 19, 1, 1–8.
422. Yong, Y., Suong, V.H. (2002) Energy approach for prediction of mechanical behavior of 2-D triaxially braided composites Part II. Parameter Analysis, *Journal of Composite Material*, 36, 10, 1233–1254.
423. Yong, Y., Suong, V.H. (2002) Energy model for prediction of mechanical behavior of 2-D triaxially braided composites, part I. Model development, *Journal of Composite Materials*, 36, 8, 963–982.
424. Zaslavskii, Y.Z., Druzhko, E.B. (1989) *New Types of Excavation Roof Bolting*, Moscow, Nedra, 295.
425. Zetterberg, T., Aström, B.T., Bäcklund, J., Burman, M. (2001) On design of joints between composite profiles for bridge deck applications, *Composite Structures*, 51, 1, 83–91.
426. Huang, Z.M. (2000) The mechanical properties of composites reinforced with woven and braided fabrics, *Composites Science and Technology*, 60, 4, 479–498.
427. Huang, Z.M., Ramakrishna, S., Thwe, A.A. (2002) Modeling and characterization of fatigue strength of laminated composites with knitted fabric reinforcement, *Journal of Composite Materials* 36, 15, 1781–1802.
428. Huang, Z.M., Ramakrishna S. (2002) Towards automatic designing of 2d biaxial woven and braided fabric reinforced composites, *Journal of Composite Materials*, 36, 13, 1541–1580.
429. Zhou, X.-F., Wagner, H.D. (2000) Fragmentation of two-fiber hybrid microcomposites: stress concentration factors and interfacial adhesion, *Composites Science and Technology*, 60, 3, 367–377.
430. Zhovner, B.A, Tsarev, V.F. (1992) Pultrusion of rectangular profiles, Aviation industry, 1, 25–26.

List of Symbols

V_S	Drawing rate
t_w	Winding pitch, stitch pitch
P_W	Winding force
t_b	Braiding strip width
n_w	Number of stitching elements
t_0	Axial braid width
n	Number of axial braids
D_{pm}	Preforming die diameter
D_d	Forming die diameter
P_{FP}	Pulling force
a	Impact viscosity
$[\tau_0]$	Limiting stresses in the braid
σ_C	Critical stress
σ_d	Stress at delamination
σ_w	Stresses in the stitch
σ_w^0	Tensile strength of sewing thread or braid
$\sigma_x^-, \sigma_y^-, \sigma_z^-$	Compressive strength in direction of corresponding axes
$\tau_{xy}, \tau_{xz}, \tau_{zy}$	Shear strength in corresponding planes
$\alpha_x, \alpha_y, \alpha_z$	Linear thermal expansion factors
k_p	Permeability factors
f_{fr}	Friction coefficient
β	Degree of binder setting
d_f	Fiber diameter
E_f	Elasticity modulus of fibers
E_b	Elasticity modulus of binder
ν_f	Poisson's ratio of fibers
ν_b	Poisson's ratio of binder
ρ	Density of reinforcing fibers
γ_w	Density of sewing material
b_s	Profile flange width
δ_s	Profile flange height

b_r	Profile post thickness
H	Profile height
B	Profile width

Index

About the Authors

Garry E. Freger is Head of Faculty for Strength of Materials at East Ukrainian National University. His research interests include: the mechanics and micromechanics of polymeric composite materials; technological process of manufacture of products and composites and carbon-carbon composites. He has published four books and 120 articles and holds patents on 45 Russian inventions. He lives in Lugansk, Ukraine.

Vladimir N. Kestelman is the author of 26 books and 516 articles on technical subjects. He holds 75 U.S. and Russian patents on his inventions. Among his books are *Electro-Physical Phenomena in the Tribology of Polymers* (Gordon and Breach, 1999); *Elecrels: Fundamentals and Applications* (Kluwer Academic, 2000); *Adhesion of Polymers* (McGraw-Hill, 2001), *Melt-Bowing Technique, Equipment, and Polymer Fibrous Materials* (Springer, 2002), and *Stress Relaxation in Viscous-Elastic Materials* (Science Publishers, 2002). He lives in King of Prussia, Pennsylvania.

Dmitry G. Freger is a degree candidate in Technical Sciences, a Scientific Employee in the Research and Design Institute at Iskra, and Senior Faculty Teacher in Strength of Materials in the Ukraine. He also lives in Lugansk.